Anitra Vickery and Mike Spooner

We c
Collaborative problem solving
for the mathematics classroom

ATM
Association of Teachers of Mathematics

Published in April 2004 by ATM

Association of Teachers of Mathematics
Unit 7 Prime Industrial Park
Shaftesbury Street
Derby DE23 8YB
Telephone: 01332 346599
Fax: 01332 204357
e-mail: admin@atm.org.uk

Copyright © 2004 Anitra Vickery, Mike Spooner

The whole of this book remains subject to copyright, but permission is granted to copy the photocopiable pages of this book for use *only* in the school which has purchased the book.

All trademarks are the property of their respective owners
All rights reserved
Printed in England

ISBN 1 898611 31 9

Copies may be purchased from the above address or http://www.atm.org.uk

Contents

We can work it out!

Collaborative problem solving for the mathematics classroom

	Introduction		4
	How to use the cards		5
1	The great race	*Logic and sequencing*	6
2	Shape arrangement	*Shape*	8
3	Football matches	*Logic and number*	10
4	What do we know? 1 – Trip to London	*Organising information–Money and time*	12
5a	Guess my number 1	*Properties of number*	15
5b	Guess my number 2	*Properties of number*	16
5c	Guess my number 3	*Properties of number*	17
6	Summer holiday	*Money*	18
7a	Mystery symbols	*Logic and number*	20
7b	Mystery symbols	*Logic and number*	21
8	Chocolate box	*Language of position*	22
9	Triangle Construction	*Angle measurement and shape*	24
10	Birthday presents	*Properties of 3D shapes*	26
11	Cake shop	*Money*	28
12	Who sits where?	*Language of position*	30
13	What do we know? 2 – Visit to the zoo	*Organising information–Money and time*	33
14	Queue for the Mirror Maze	*Logic and sequencing*	35
15a	Elimination game	*Money and properties of number*	37
15b	Elimination game	*Money and properties of number*	38
15c	Elimination game	*Money and properties of number*	40
16	Inscribed shape	*Properties of shape*	42
17	Triangle	*Shape and number*	44
18	The culprit	*Logic and sequencing*	45
19	Pie chart	*Data handling, time, fractions*	48
20	Flight times	*Timetables*	51
21	Who lives where?	*Logic and deduction*	53
22	Solid shapes	*Properties of 3D shape*	56
23	School trip	*Money, number and time*	58
24	Framework shapes	*Properties of 2D and 3D shape*	61
25	Book shop	*Money, percentage increases and discounts*	63
	Answers		66

We can work it out!
Collaborative problem solving for the mathematics classroom

What are group problem-solving cards?

The pack contains 25 problem-solving activities. Each activity is presented so that it can be cut up to make a collection of cards. Each set of cards poses a problem and contains all the information needed to solve it. The sets also contain information that is correct but of no relevance or use in solving the problem. If the activities are used according to the instructions they will also fulfil the requirement for speaking and listening across the curriculum.

Has the resource been tried out with children?

The cards have been introduced to teachers in INSET sessions. Teachers from over 80 schools have provided feedback and the authors have carried out close observation of children working with the cards in two schools.

What were the results of the research?

A number of benefits were noted by teachers in using these activities, including the following:

- They are highly motivational.
- Groups with a wide range of attainment can be active participants.
- They put problem solving into meaningful or interesting contexts.
- They teach children not to expect immediate answers to problems.
- They make multi-step problems more manageable and in so doing encourage a strategic approach.
- They encourage children to organise information and identify redundant information.
- They develop skills of collaboration and cooperation.
- They encourage children to check their answers.

Can the skills used in working with the cards be transferred?

Teachers who took part in the trials of the approach reported that children began to apply some of the skills and behaviours they exhibited when working with the cards to other work. In particular, they reported a more logical and strategic approach to organising information and a greater inclination to check answers. This is more likely to happen when there is an explicit analysis of the approach that was followed to reach a solution.

Which teaching objectives are addressed by the activities?

The activities draw on a range of different areas of mathematics including number, shape, time, data handling and money as well as focussing on more general thinking skills such as logical deduction and the organisation of information. The problems use vocabulary that meets the requirements of the National Curriculum and the National Numeracy Strategy.

Some exercises emphasise skills and concepts that are related to particular mathematical themes:

Mathematical themes	Exercises
Properties of numbers	5 15 17
Fractions and decimals	3 23 25
Calculation	6 13 23 25
Reasoning about numbers	3 4 15
Reasoning about shape	2 8 9 10 12 16 17 22 24
Problems involving money	4 6 11 13 15 23
Problems involving measures	4 13 20 23
Handling data	19
Shape and space	2 8 9 10 16 17 22 24
Measures	4 13 20

We can work it out!

How to use the cards

Which children is the resource aimed at?

The activities have been written for children working from level 3 to level 6 of the National Curriculum in mathematics (that is to say, upper primary and lower secondary). When the cards are used as suggested below the collaborative nature of the exercise enables children with a wide range of attainment to take part and experience success. However, some sets of cards will be more appropriate than others for different age groups.

About the Cards.

The sets of cards are grouped in ascending order of difficulty. Most sets of cards are group problem solving activities and are presented in a similar format, with the question to be answered forming part of the set of cards, and also some 'red herrings' in each set. Each set of cards poses a problem and contains all the information needed to solve it. However, four sets of activities are slightly different.

- Activities 4 and 13, entitled 'What do we Know 1 & 2' ask the children to sort the cards into three groups. Questions for which they know the answers, questions needing more information for an answer, and questions that cannot be answered at all.

- Activities 5 and 15 encourage the children to eliminate numbers using a series of clues until only one number remains.

How to use the cards

To realise the full potential of all the activities we suggest the teacher models the activity as outlined below. Initially it may be helpful, for example, to use the same problem with all groups in the class.

- The title card for each exercise records the number of cards there are in each set (excluding the title card) and it is important to check that the set is complete before starting.

- For some activities such as 7, 11, 15a and b and 20, it may also be useful to photocopy and laminate the relevant resources beforehand.

- Organise the children to work in mixed attainment groups of about four.

- Share out the cards between the members of each group.

- Ask the children to:
  - read through the cards;
  - find the one that describes the objective of the exercise;
  - organise the cards, e.g. identify essential or redundant information or information that refers to particular aspects of the problem.

- Work through the information encouraging children to use resources as appropriate e.g. coloured objects, name cards etc. This promotes the kinaesthetic aspects of the activity.

- Encourage the children to check their solution and to feed back about the process they followed, about what helped them and about what obstructed them.

- Once the children are familiar with the approach they could be allowed to choose which problems to work on and encouraged to extend their choice when necessary or to construct their own exercises.

- The approach can be extended for use with mathematical problems presented in a more traditional format and with real-life problems.

The great race

Activity 1

14 cards

Find the finishing order of the cars

There are nine cars in the race

All the cars finished the race

The white car finished six places behind the blue car

The orange car finished between the green and the yellow cars

---

It was raining on the day of the race

Only one car finished ahead of the blue car

The yellow car was in the middle of the order of finishing

The purple car finished behind the white car

The green car finished ahead of the black car

The pink car finished between the yellow and black cars

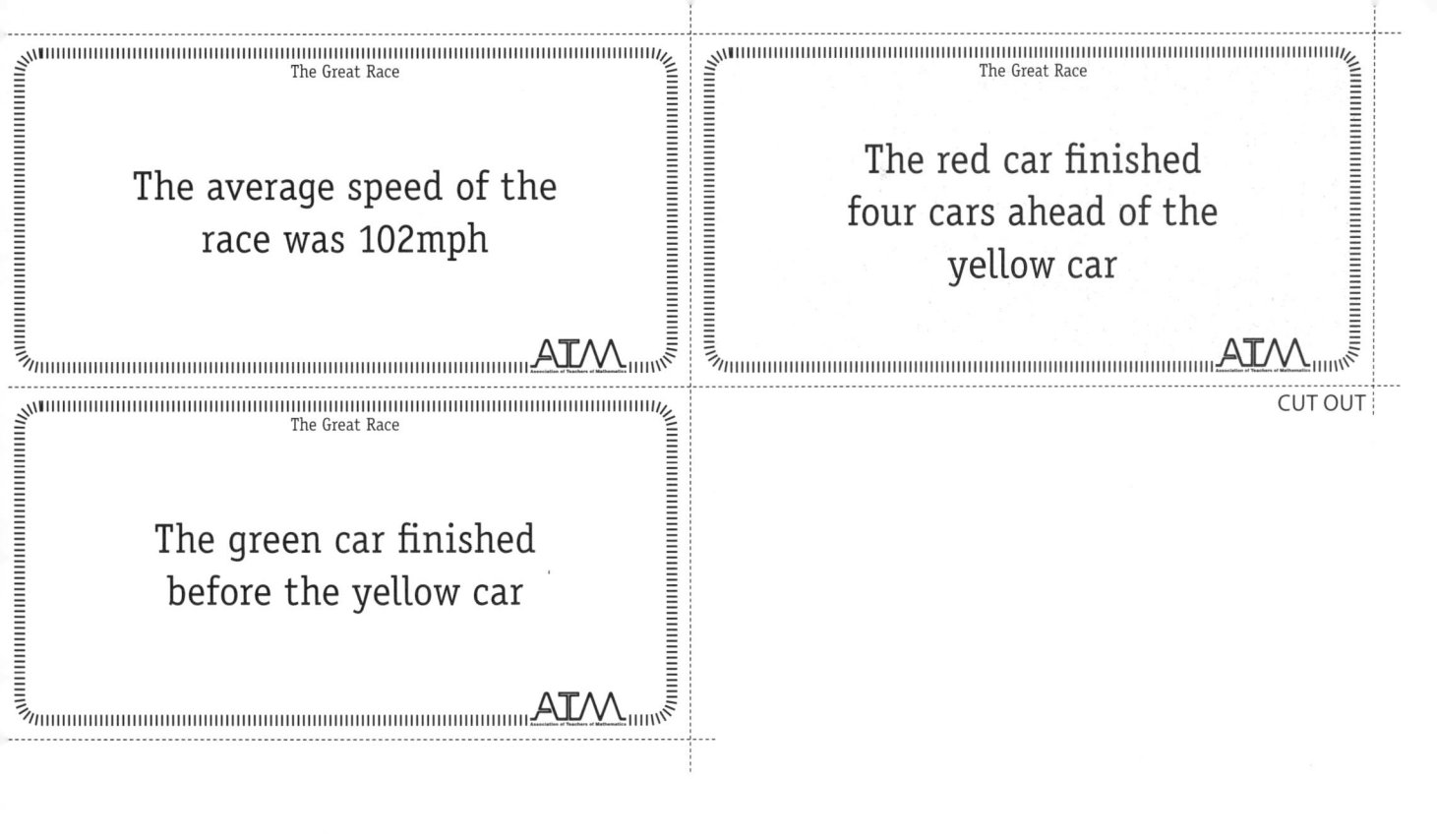

The Great Race

The average speed of the race was 102mph

The Great Race

The red car finished four cars ahead of the yellow car

The Great Race

The green car finished before the yellow car

Shape arrangement

Activity 2

15 cards

Shape arrangement

Discover the colour of each shape and how they have been arranged

---

Shape arrangement

All the shapes with straight sides are regular shapes

Shape arrangement

There are two rows of 4 shapes

Shape arrangement

Yellow and blue shapes are next to each other in the top row

Shape arrangement

The shapes in the top row are the same as the shapes in the bottom row but in the reverse order

---

Shape arrangement

The blue shape has no straight lines

Shape arrangement

There are four different shapes in the arrangements

Shape arrangement

The shape with the largest number of sides is green

Shape arrangement

The hexagon on the bottom row is below a triangle

Shape arrangement

The first shape in the top row is half the area of the second shape

---

Shape arrangement

Each different shape has its own colour

The number of letters in the colour of the shape with the smallest number of sides is the same as the number of sides

One of the shapes has 4 sides that are equal

What are the names of the shapes in the arrangement?

The blue shape has a diameter that is equal to the length of one side of the yellow shape

CUT OUT

Football Matches

Activity 3

20 cards

The biggest crowd of the day was at Manchester United's stadium

---

Manchester United were the only team that did not score

Chelsea scored their first goal in the 10th minute

There were 5 000 away supporters at the Chelsea stadium

Liverpool scored twice as many goals as Chelsea

There were 4 500 Southampton supporters at Liverpool's ground

---

Arsenal scored three goals

All the tickets for Liverpool's game were sold out

Use the clues to find out the teams that played, the scores and attendances at four football matches

---

Only 10% of the crowd at Liverpool were away supporters

Football Matches

- Chelsea scored their second goal just before the referee blew the final whistle

- There were 6 000 fewer people at Chelsea's ground than Liverpool's

- All of the away teams except Arsenal scored just one goal

- The smallest crowd was at Wolverhampton and it was 3 000 fewer than the crowd at Chelsea

- Arsenal had no goals scored against them

- There were 14 000 more people in the crowd at Manchester than in the 2nd biggest crowd

- Only 5 000 Spurs supporters saw their team play

- Southampton had the most goals scored against them

- In their match Wolves and Newcastle did not win or lose

- Arsenal scored more goals than Chelsea but fewer than Liverpool

CUT OUT

11

What do we know? 1
Trip to London
Activity 4

10 question cards *Q*
17 information cards *i*

CUT OUT AND LAMINATE

Instructions

Separate the *Q* Question cards into three groups
a) Questions which you can answer
b) Questions which you can't answer accurately because you need more information
c) Questions you can't answer at all.
Give the answers when you can or when you can't, explain what else you would need to know.

Q 1. How old is John?

Q 2. How much did John spend on the train?

Q 3. How long did John stay in London?

Q 4. How long did the train journey take?

Q 5. How many miles was the train from London when it was delayed?

Q 6. How much did John pay for his ticket?

Q 7. At what time was John's appointment with the queen?

Q 8. How much did John save on his ticket because of his senior citizen discount?

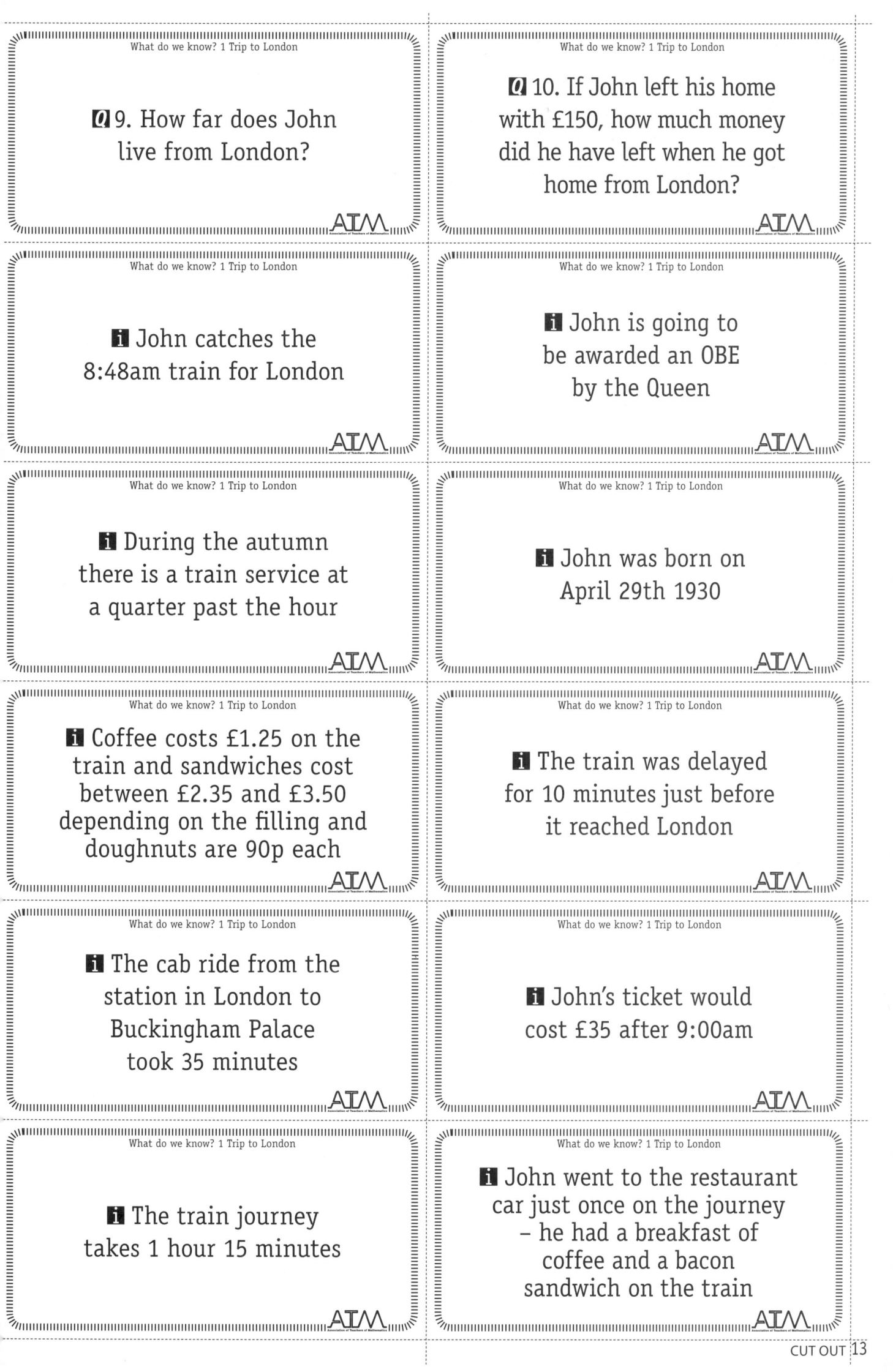

What do we know? 1 Trip to London

Q 9. How far does John live from London?

What do we know? 1 Trip to London

Q 10. If John left his home with £150, how much money did he have left when he got home from London?

What do we know? 1 Trip to London

i John catches the 8:48am train for London

What do we know? 1 Trip to London

i John is going to be awarded an OBE by the Queen

What do we know? 1 Trip to London

i During the autumn there is a train service at a quarter past the hour

---

What do we know? 1 Trip to London

i John was born on April 29th 1930

What do we know? 1 Trip to London

i Coffee costs £1.25 on the train and sandwiches cost between £2.35 and £3.50 depending on the filling and doughnuts are 90p each

---

What do we know? 1 Trip to London

i The train was delayed for 10 minutes just before it reached London

---

What do we know? 1 Trip to London

i The cab ride from the station in London to Buckingham Palace took 35 minutes

---

What do we know? 1 Trip to London

i John's ticket would cost £35 after 9:00am

What do we know? 1 Trip to London

i The train journey takes 1 hour 15 minutes

What do we know? 1 Trip to London

i John went to the restaurant car just once on the journey – he had a breakfast of coffee and a bacon sandwich on the train

CUT OUT 13

What do we know? 1 Trip to London

i After receiving his medal, John had a good time celebrating in London

i A child's fare is half the adult fare

i Senior citizens are eligible for a 20% discount on all train fares

i John gets to the palace just in time for his appointment

i Adult tickets cost £70 before 9:00am

i The train company gives a senior citizen discount to everyone who is over 60 years old

i Coffee is free if the train is more than 30 minutes late

Guess my number 1 Find the number between 1 and 99	**Guess my number 1** **Activity 5a** **7 cards**
Guess my number 1 The digital sum is 6	Guess my number 1 One of the digits is a 2
Guess my number 1 It is more than 5 squared	Guess my number 1 It is less than 55
Guess my number 1 The number is a multiple of 3	Guess my number 1 It is not a square number

CUT OUT

Guess my number 2

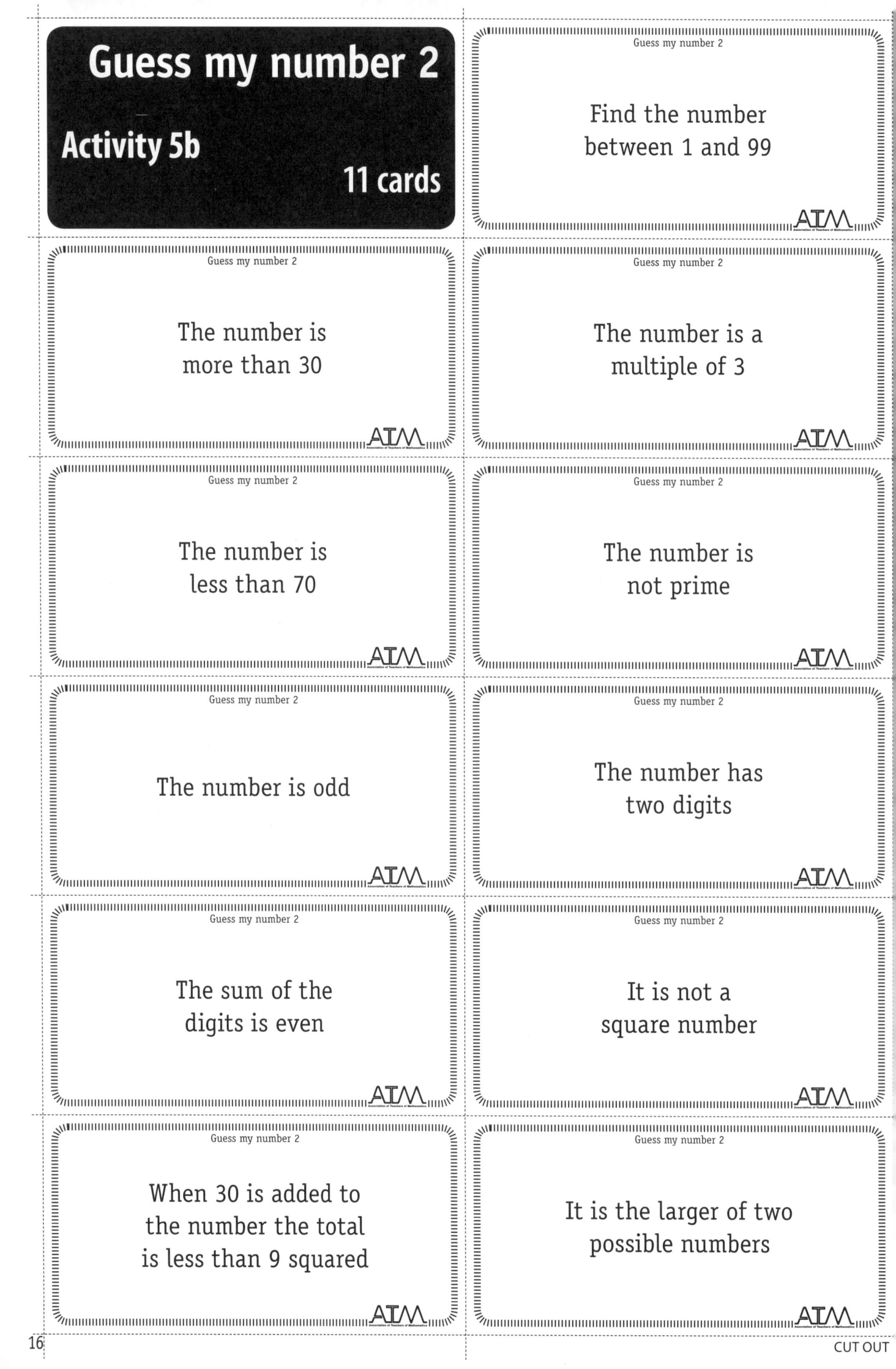

Activity 5b — 11 cards

Find the number between 1 and 99

The number is more than 30

The number is a multiple of 3

The number is less than 70

The number is not prime

The number is odd

The number has two digits

The sum of the digits is even

It is not a square number

When 30 is added to the number the total is less than 9 squared

---

It is the larger of two possible numbers

Guess my number 3

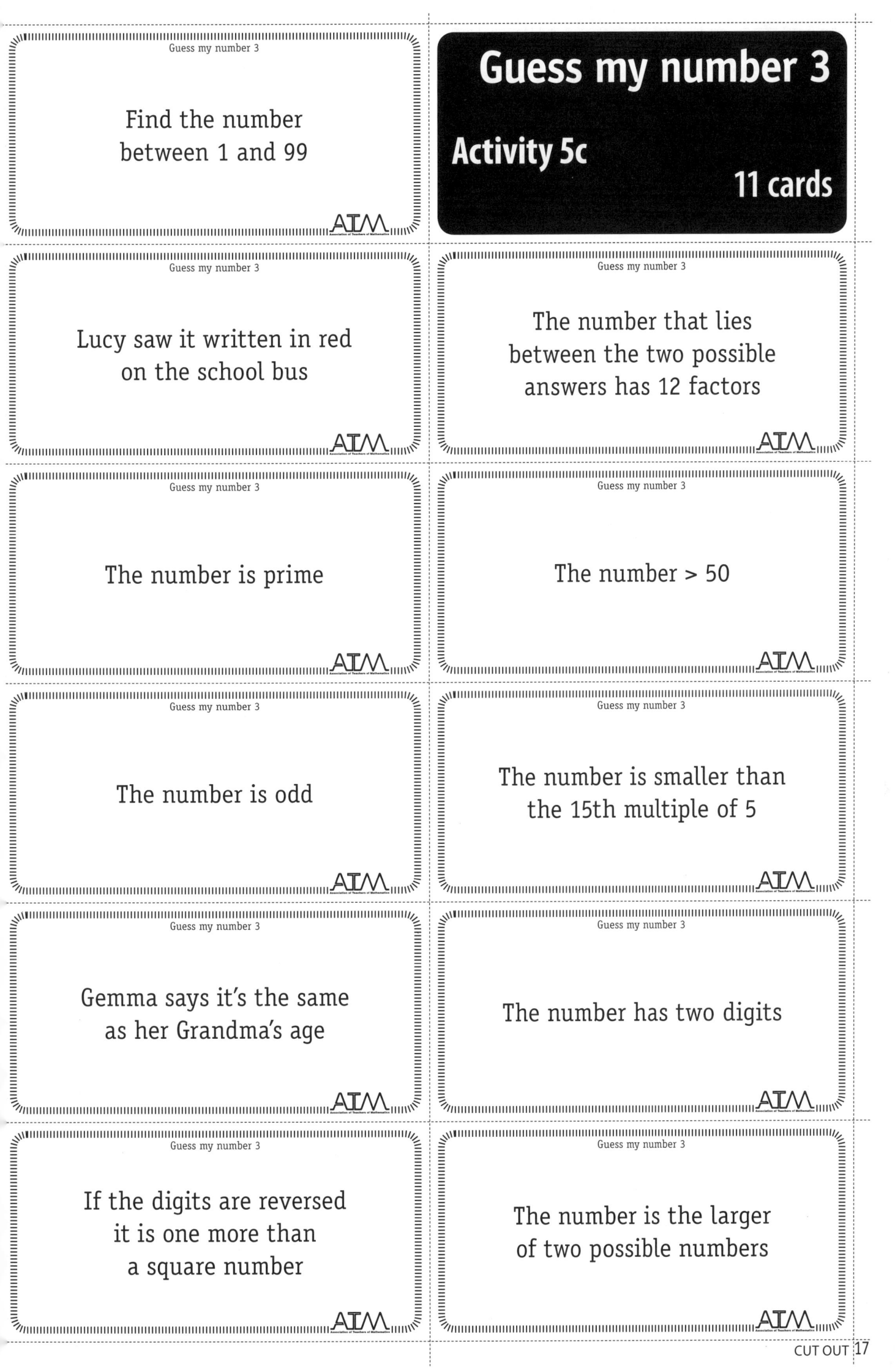

Activity 5c — 11 cards

- Find the number between 1 and 99

- Lucy saw it written in red on the school bus

- The number that lies between the two possible answers has 12 factors

- The number is prime

- The number > 50

- The number is odd

- The number is smaller than the 15th multiple of 5

- Gemma says it's the same as her Grandma's age

- The number has two digits

- If the digits are reversed it is one more than a square number

- The number is the larger of two possible numbers

Summer Holiday

Activity 6

14 cards

Summer Holiday

The Banks family decides to go to Spain for a holiday – work out how much the trip costs, including spending money and car hire

---

Summer Holiday

Mr Banks is 3 years younger than Mrs Banks

Summer Holiday

Baby John's age in months is the same as Jodie's and Harry's combined age in years

---

Summer Holiday

A two-week package with accommodation, flights and meals costs £345 for each adult

---

Summer Holiday

Children under two years of age go free

Summer Holiday

There are 5 people in the Banks family; Mum and Dad, Jodie who is 12, Harry who is 7 and baby John

---

Summer Holiday

Mr Banks decides to hire a car for ten days

Summer Holiday

The cost of the holiday for each child is £250

Summer Holiday

Car hire for one week costs £170

Summer Holiday

The family decide to take £600 spending money

Summer Holiday

The daily rate for car hire is £36

CUT OUT

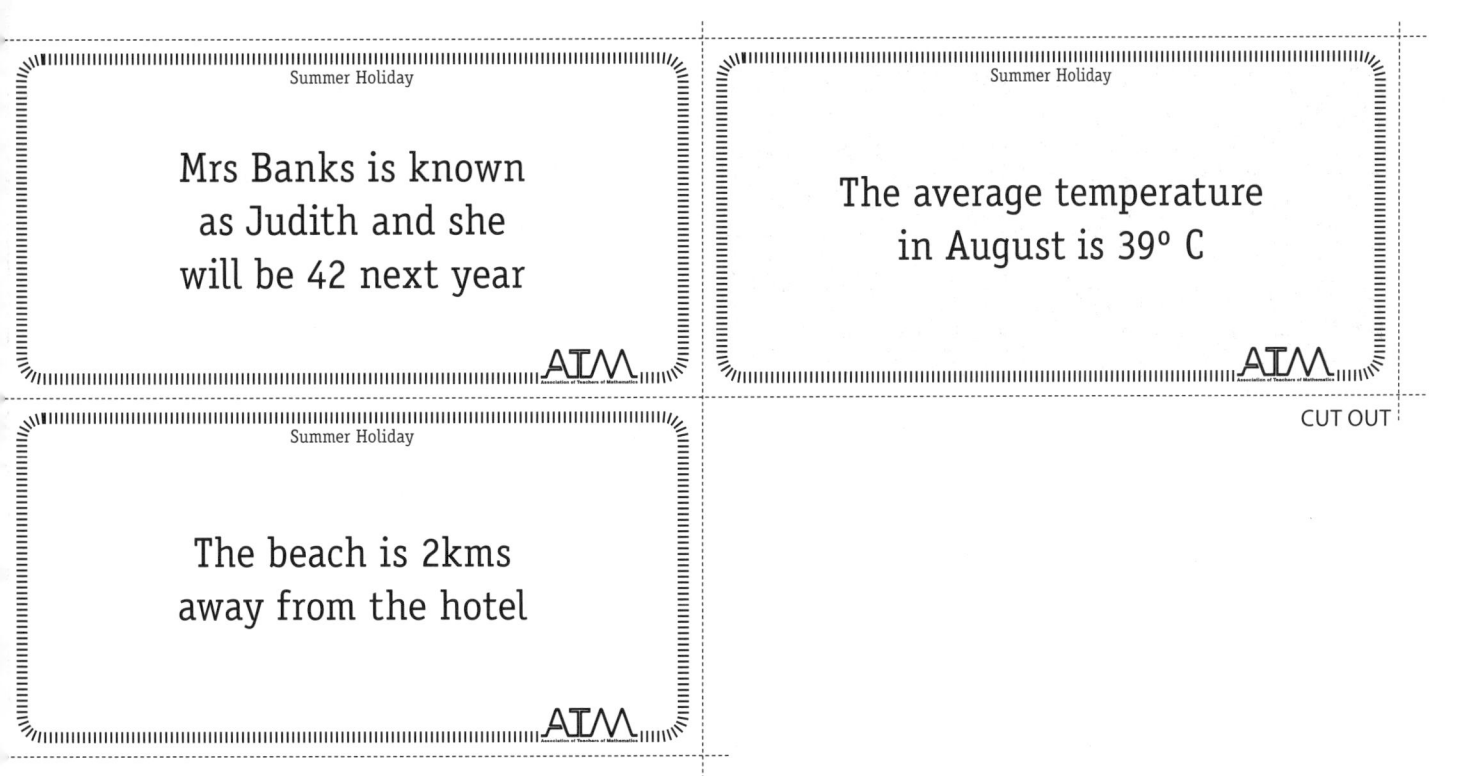

Summer Holiday

Mrs Banks is known as Judith and she will be 42 next year

Summer Holiday

The average temperature in August is 39° C

Summer Holiday

The beach is 2kms away from the hotel

Mystery symbols

Activity 7a — 1 card

Mystery symbols 7a

Each symbol represents a different digit (1-9) – find out which digit is represented by each symbol

Mystery symbols 7a

★	★	★	★	8
£	$	e	♣	17
★	♠	★	♠	16
$	★	€	♣	11
9	11	14	18	

CUT OUT AND LAMINATE

Mystery symbols

Activity 7b

8 cards

Mystery symbols 7b

Each symbol represents a different digit (1-9) – find out which digit is represented by each symbol

---

Mystery symbols 7b

☐ + ☐ = ✺

Mystery symbols 7b

▲ − ♦ = ☐

Mystery symbols 7b

✺ + ✺ = ✺ × ✺

Mystery symbols 7b

❀ − ☐ = Ω

Mystery symbols 7b

✎ × ✎ = ♥

Mystery symbols 7b

✺^{3} = ❀

Mystery symbols 7b

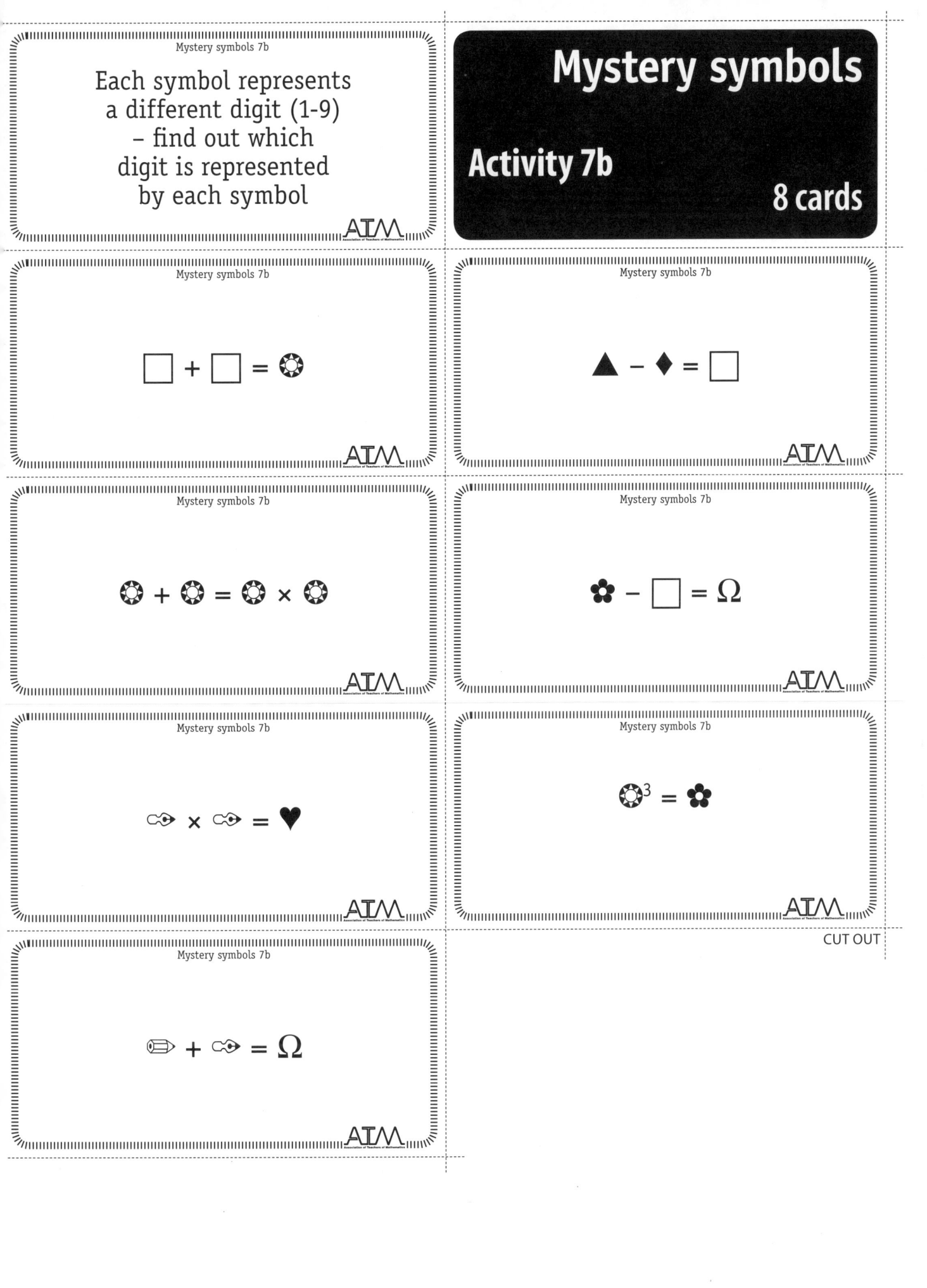

✐ + ✎ = Ω

Chocolate Box

Activity 8 — 15 cards

Draw a plan view of the chocolates in a box

The second row has 3 chocolates and the third just 2

The space for the Lemon Twist touches the bottom of the space for Turkish Delight

---

Delightful Chocolates are sold in triangular boxes that have just one layer

---

The two chocolates with nuts are next to each other

There are ten chocolates altogether

There is a mixture of dark and light chocolates

There are four rows of chocolates in the box

The Truffle Delight is in a row on its own

The first row is at the base of the triangle and has four chocolates in it

---

The Creamy Caramel is in the centre of a row

CUT OUT

Chocolate Box

Strawberry Fizz and Lemon Twist are in the same row as Creamy Caramel

Chocolate Box

Coconut Heaven is in the row that has half the number of chocolates as the bottom row

Chocolate Box

The spaces for Truffle Delight, Turkish Delight, Lemon Twist and Hazelnut Crackle all touch the right hand side of the box

Chocolate Box

Liquid Cherry lies between Tantalising Cream and Chocolate Brazil

CUT OUT

Triangle Construction

Activity 9 — 11 cards

Use the information gained from the clues to make an accurate drawing of the triangle

---

The number of degrees in the obtuse angle is included in a hundred square

---

The number of degrees in the largest angle is divisible by 3

The triangle is symmetrical

If the digits of the number of degrees of the largest angle are reversed it is divisible by 3

The triangle is isosceles

Each angle has a whole number of degrees

The triangle can be drawn in any colour

The number of degrees in the largest angle has four more factors than the number of degrees in the smaller angles

---

The length, in centimetres, of the side opposite the obtuse angle is 1/6th of the number of degrees in one of the other angles

---

All the angles are multiples of 6

CUT OUT

Birthday presents

Activity 10

22 cards

Find out the shape of the parcels and work out which present was in each parcel

---

Sophie received 6 parcels for her birthday

The parcel for the Maltesers® had three surfaces, two flat and one curved

---

The number of faces in the net of the parcel for the hair decorations was a factor of the number of hair decorations

---

Two of the parcels were the same shape as the presents they contained

---

The circumference of the silver bracelet was almost the same as the circumference of the only flat face in its parcel

---

All the parcels were three dimensional

The net of the parcel containing hair decorations had 2 different shapes, six of one and two of the other

---

The parcel for the bracelet had a vertex

Sophie always asks for Maltesers® for her birthday

Four of the parcel shapes had the same initial letter

Birthday presents

Sophie wanted 16 different hair decorations

Birthday presents

There were 72 Maltesers® in the tube

Birthday presents

Two of the parcels contained at least one circular face

Birthday presents

The bubble bath made a "slopping" noise when you rolled the parcel

Birthday presents

The parcel for the bracelet had two surfaces, one flat and one curved

Birthday presents

Sophie knew what presents she was going to get

Birthday presents

One of the parcels had only one surface

Birthday presents

Her fluffy jumper was contained in a parcel in which the three dimensions were equal

Birthday presents

The number of surfaces on each parcel was a factor of the number of Maltesers®

Birthday presents

All of the parcels were different solids

Birthday presents

Sophie asked for the latest Harry Potter book

CUT OUT

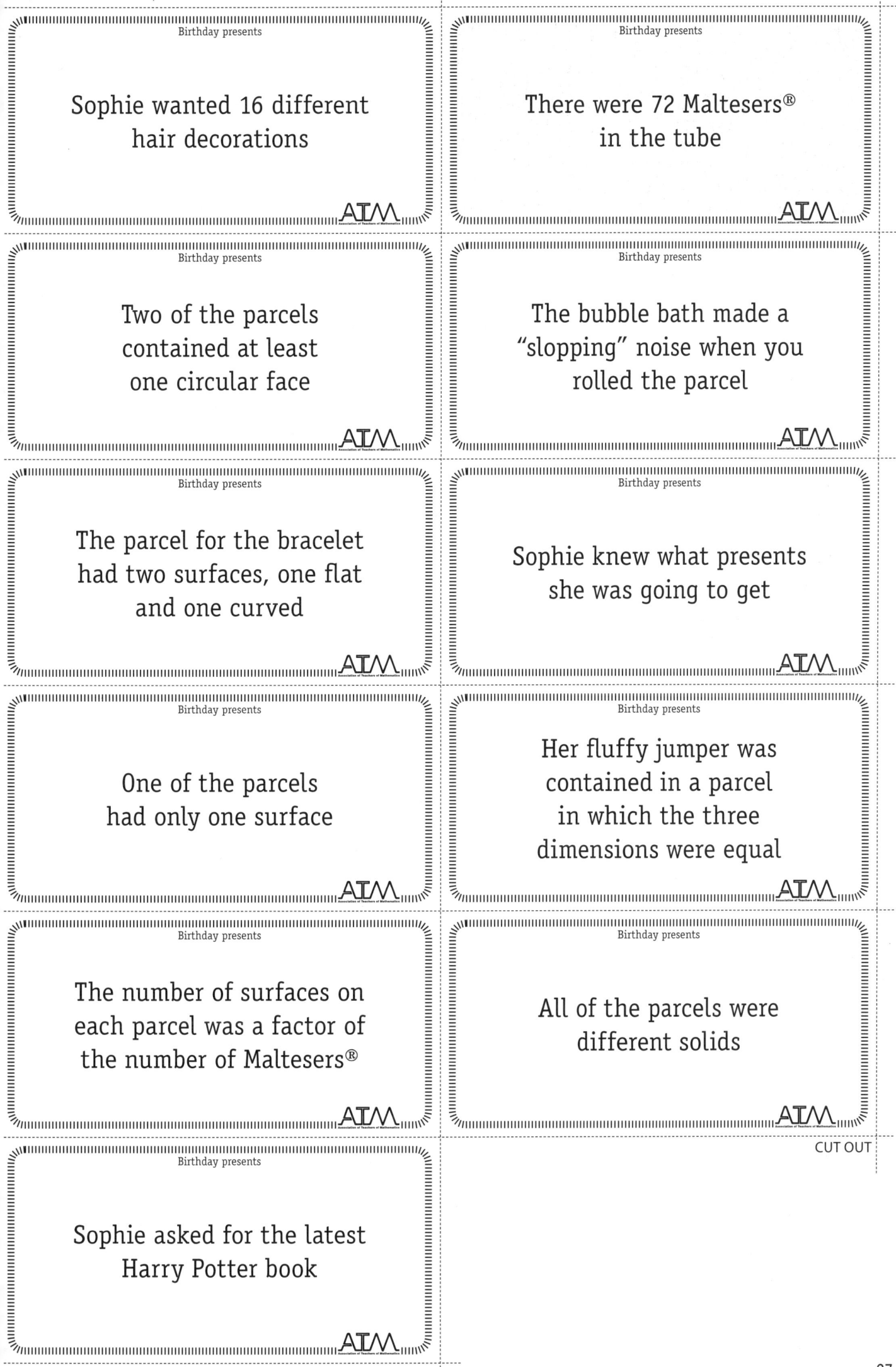

27

Cake shop

Activity 11 — 19 cards

Cake shop

Complete the price list for the cakes

Cake shop

Price List

Name of Cake		Price
	A	
	B	
	C	
	D	
	E	

Cake shop

A cake shop sells 5 different sorts of cakes (A, B, C, D and E) – use the information to find out what type of cakes they are and their prices

---

Cake shop

Cake A is a bun with currants in it that has the same name as a football team from London

---

Cake shop

Two of the prices are odd numbers

Cake shop

You could pay for the most expensive cake with a large heptagonal coin and you would get no change

---

Cake shop

All the prices are multiples of 5

Cake shop

Cake B is brown, covered in sugar and has jam in it

Cake shop

The cheapest cake is exactly half the price of the most expensive

---

Cake shop

The price of Cake B could be paid exactly with three different silver coins

---

CUT OUT

Cake shop

If you bought one of each of the buns it would cost £1.80

Cake shop

Cake C is traditionally eaten on Good Friday

Cake shop

The price of Cake E could be paid exactly with two of the same silver coin

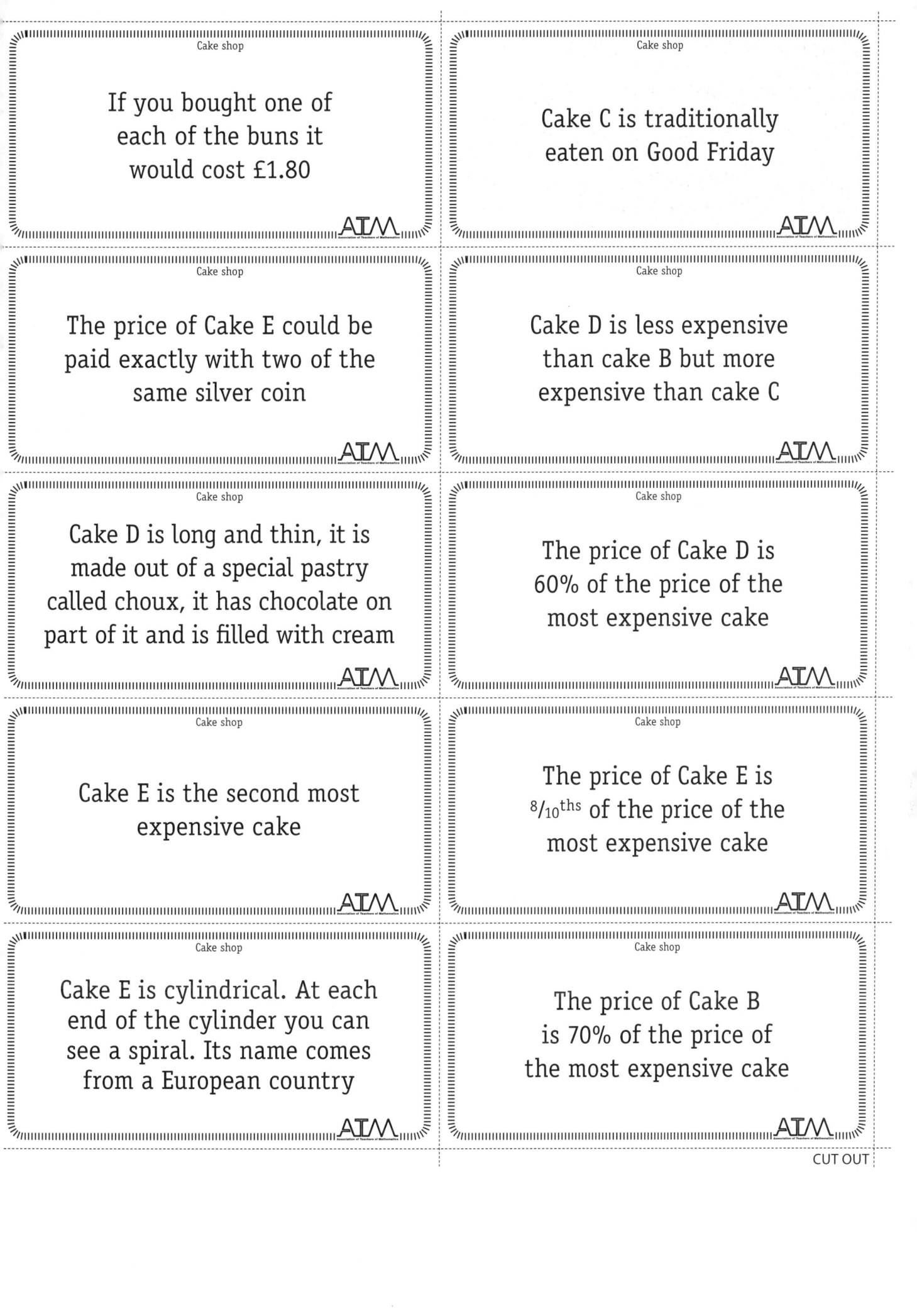

Cake shop

Cake D is less expensive than cake B but more expensive than cake C

Cake shop

Cake D is long and thin, it is made out of a special pastry called choux, it has chocolate on part of it and is filled with cream

---

Cake shop

The price of Cake D is 60% of the price of the most expensive cake

---

Cake shop

Cake E is the second most expensive cake

Cake shop

The price of Cake E is $8/10$ths of the price of the most expensive cake

---

Cake shop

Cake E is cylindrical. At each end of the cylinder you can see a spiral. Its name comes from a European country

---

Cake shop

The price of Cake B is 70% of the price of the most expensive cake

Who sits where?

Activity 12

28 cards

Who sits where?

Work out where each child sits

Who sits where?

Board

Who sits where?

Barry has only 3 other people at his table

Who sits where?

Anita shares a table with Ben, Ahmed, Eloise, Jim and Megan

Who sits where?

Holly shares a table with Deepak, Diego and Louisa

Who sits where?

Holly sits at the desk that is nearest to one of the back corners of the room

---

Who sits where?

Catalina shares a table with Barry, Ranjit and Satvia

Who sits where?

Alex shares a table with Julie, George, Patrick, Sophia and Harshini

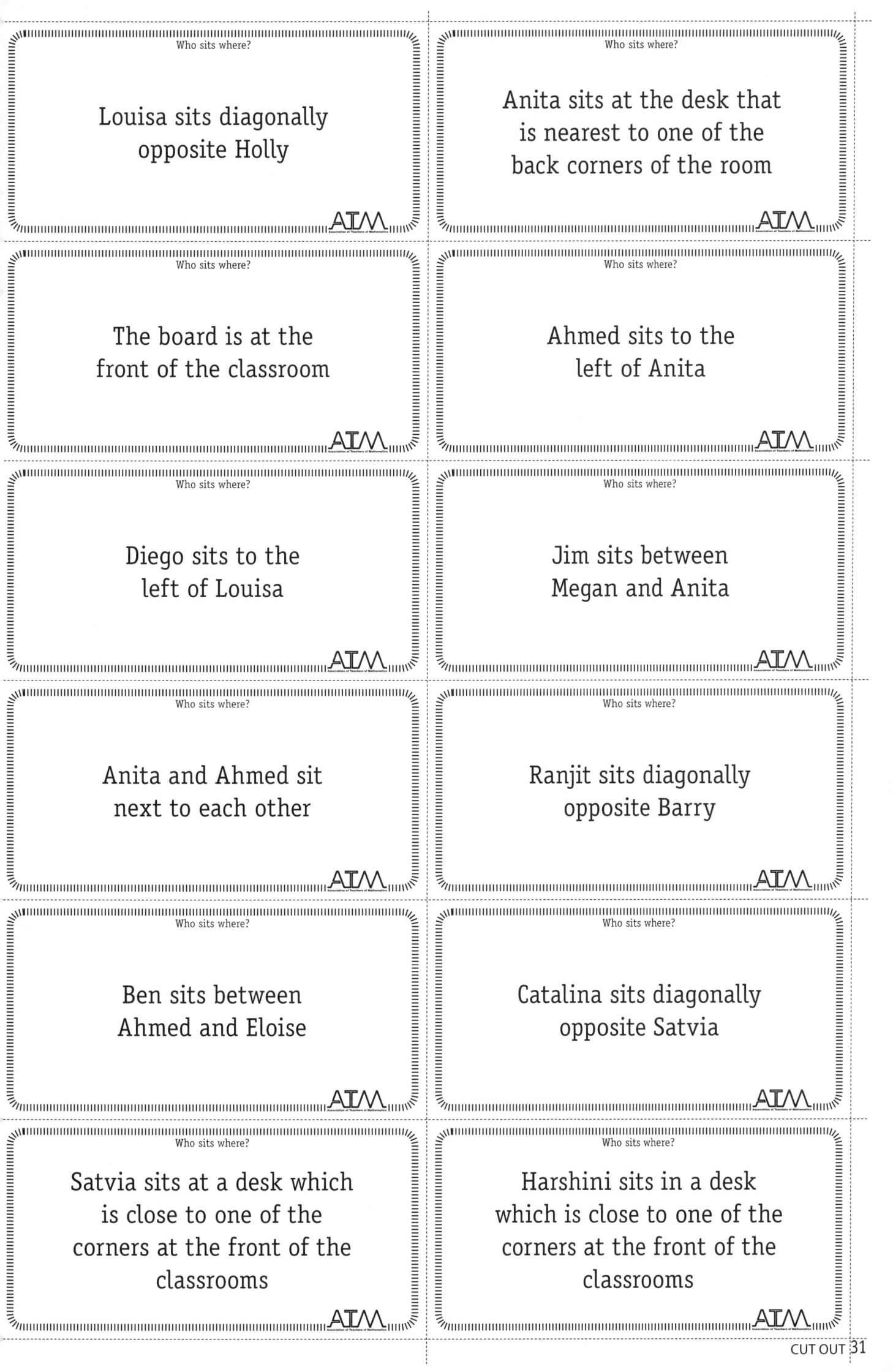

Who sits where?
Louisa sits diagonally opposite Holly

Who sits where?
Anita sits at the desk that is nearest to one of the back corners of the room

Who sits where?
The board is at the front of the classroom

Who sits where?
Ahmed sits to the left of Anita

Who sits where?
Diego sits to the left of Louisa

Who sits where?
Jim sits between Megan and Anita

Who sits where?
Anita and Ahmed sit next to each other

Who sits where?
Ranjit sits diagonally opposite Barry

Who sits where?
Ben sits between Ahmed and Eloise

Who sits where?
Catalina sits diagonally opposite Satvia

Who sits where?
Satvia sits at a desk which is close to one of the corners at the front of the classrooms

Who sits where?
Harshini sits in a desk which is close to one of the corners at the front of the classrooms

CUT OUT 31

Who sits where?

Catalina sits next to Ranjit

Who sits where?

George sits between Sophia and Alex

Who sits where?

Satvia sits nearer to the front of the classroom than Barry

Who sits where?

George sits facing the board

Who sits where?

Patrick sits between Harshini and Julie

Who sits where?

Harshini sits at the furthest end of a diagonal from Sophia

Who sits where?

George sits opposite Patrick

Who sits where?

Alex sits at the furthest end of a diagonal from Julie

Who sits where?

Patrick is nearer to the front of the room than George

CUT OUT

i The driver will be Eleanor's father, Fred

What do we know? 2
Visit to the Zoo
Activity 13

10 question cards *Q*
11 information cards *i*

Instructions

Separate the *Q* Question cards into three groups

a) Questions which you can answer
b) Questions which you can't answer accurately because you need more information
c) Questions you can't answer at all

Give the answers when you can or when you can't, explain what else you would need to know?

Q 2. What is the favourite animal of the children in the class?

Q 7. How much will it cost altogether for the group to go into the zoo?

Q 3. What time will they leave the zoo?

Q 8. How far away from the school is the zoo?

Q 4. How many adults will be on the bus?

Q 9. If there was one more child in the class, how much would they pay altogether for the class to go into the zoo?

Q 5. How much will it cost for each child to go into the zoo?

Q 10. What time will they arrive at the zoo?

CUT OUT 33

What do we know? 2 Visit to the Zoo

Q 1. At what time will they start to feed the seals?

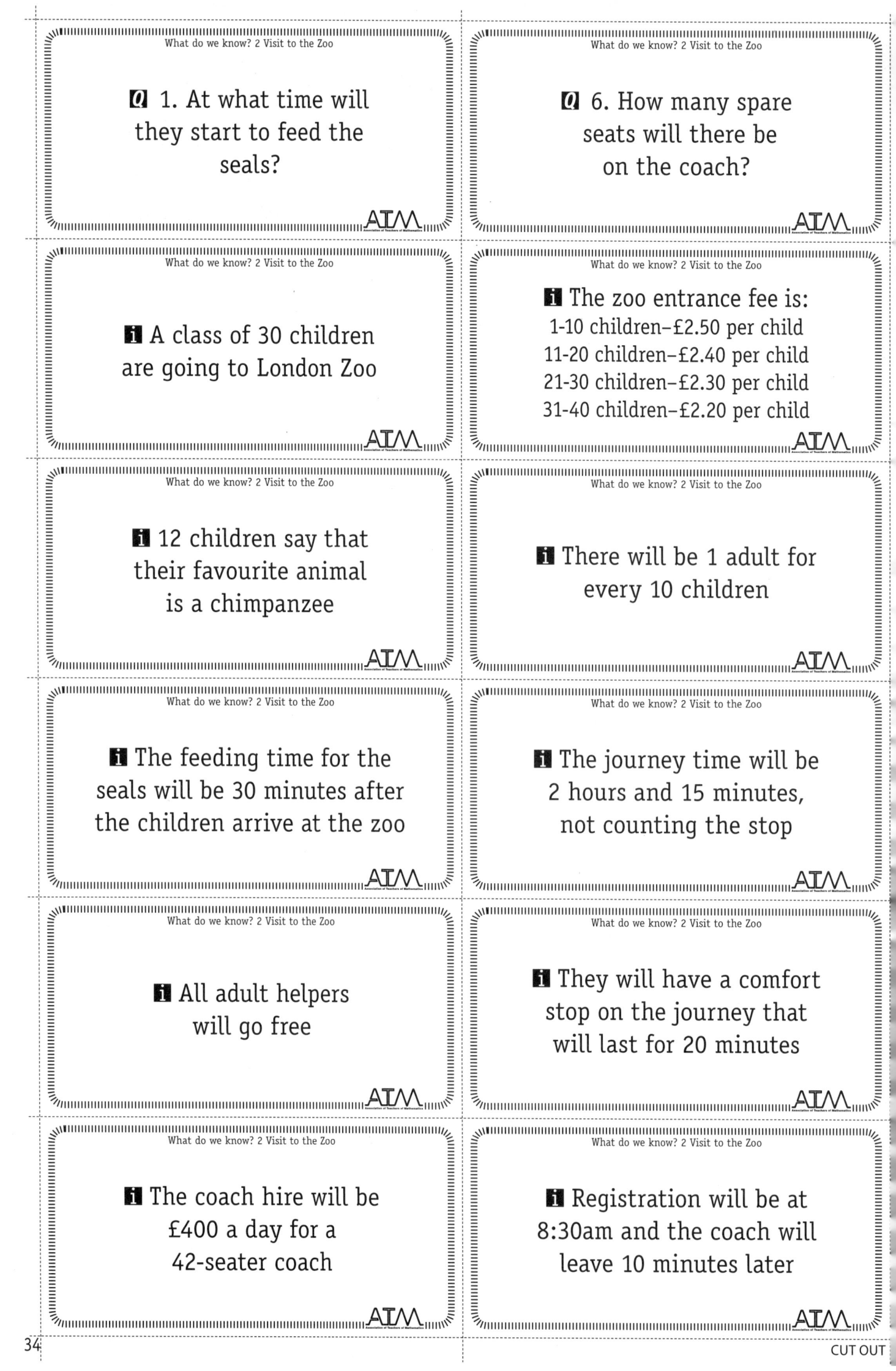

Q 6. How many spare seats will there be on the coach?

---

i A class of 30 children are going to London Zoo

---

i The zoo entrance fee is:
1-10 children – £2.50 per child
11-20 children – £2.40 per child
21-30 children – £2.30 per child
31-40 children – £2.20 per child

---

i 12 children say that their favourite animal is a chimpanzee

---

i There will be 1 adult for every 10 children

---

i The feeding time for the seals will be 30 minutes after the children arrive at the zoo

---

i The journey time will be 2 hours and 15 minutes, not counting the stop

---

i All adult helpers will go free

---

i They will have a comfort stop on the journey that will last for 20 minutes

---

i The coach hire will be £400 a day for a 42-seater coach

---

i Registration will be at 8:30am and the coach will leave 10 minutes later

Queue for the Mirror Maze
Activity 14
21 cards

Find the order of the first queue for the Mirror Maze

Pippa and Sarah are twins but Pippa was born 1 hour before Sarah

---

The children of class 7 went to Adventure Land for their school trip

---

Satvia has a brother who is doctor

A girl was the first in the queue

The oldest child went first

Ahmed was the youngest in the first group by a whole month

Grace is only 1 day older than Pippa

Sarah had just quarrelled with Ahmed so she was glad she wasn't standing next to him

---

Andrew was wearing black trainers

Pippa and James liked each other and kept holding hands when they were standing in the queue

CUT OUT 35

Queue for the Mirror Maze

John and Ahmed were best friends and were cross because they were the furthest apart possible

---

Satvia and Andrew share the same birthday

Ahmed's birthday is on July 29th

Mr Stevens told Mike off for pushing and shouting

Mr Stevens would have to organise a queue three times

Ahmed had 56p of his spending money left by the time they got to the Mirror Maze

---

Sally was very ill when she was 5 and stayed back a year in school

---

The back of Grace's white shoes were spoiled by Mike's muddy boots whilst she was standing in the queue

---

Only 10 pupils at a time were allowed into the Mirror Maze

Mr Stevens, the class teacher, chose the 5 oldest girls and the 5 oldest boys for the first queue and lined them up alternately (girl-boy-girl-boy)

CUT OUT

Elimination Game a — Activity 15a — 9 cards

25p	£1.12	56p	12p	£5.25
45p	20p	£2	78p	£30
90p	16p	85p	£34.40	£10.50
£3.33	£27.75	38p	£16	£2.55
87p	£125.45	5p	£45	£21.98

Use the clues to find the sum of money

The amount cannot be made with just notes

The amount cannot be made with just one coin

The amount is less than the value of the least valuable note

The amount is less than the value of the most valuable coin

The amount can be made by using just silver coins

The amount is greater than the value of the most valuable silver coin

---

You must use coins of at least two different values to make the amount

---

It is the lesser of two possible amounts

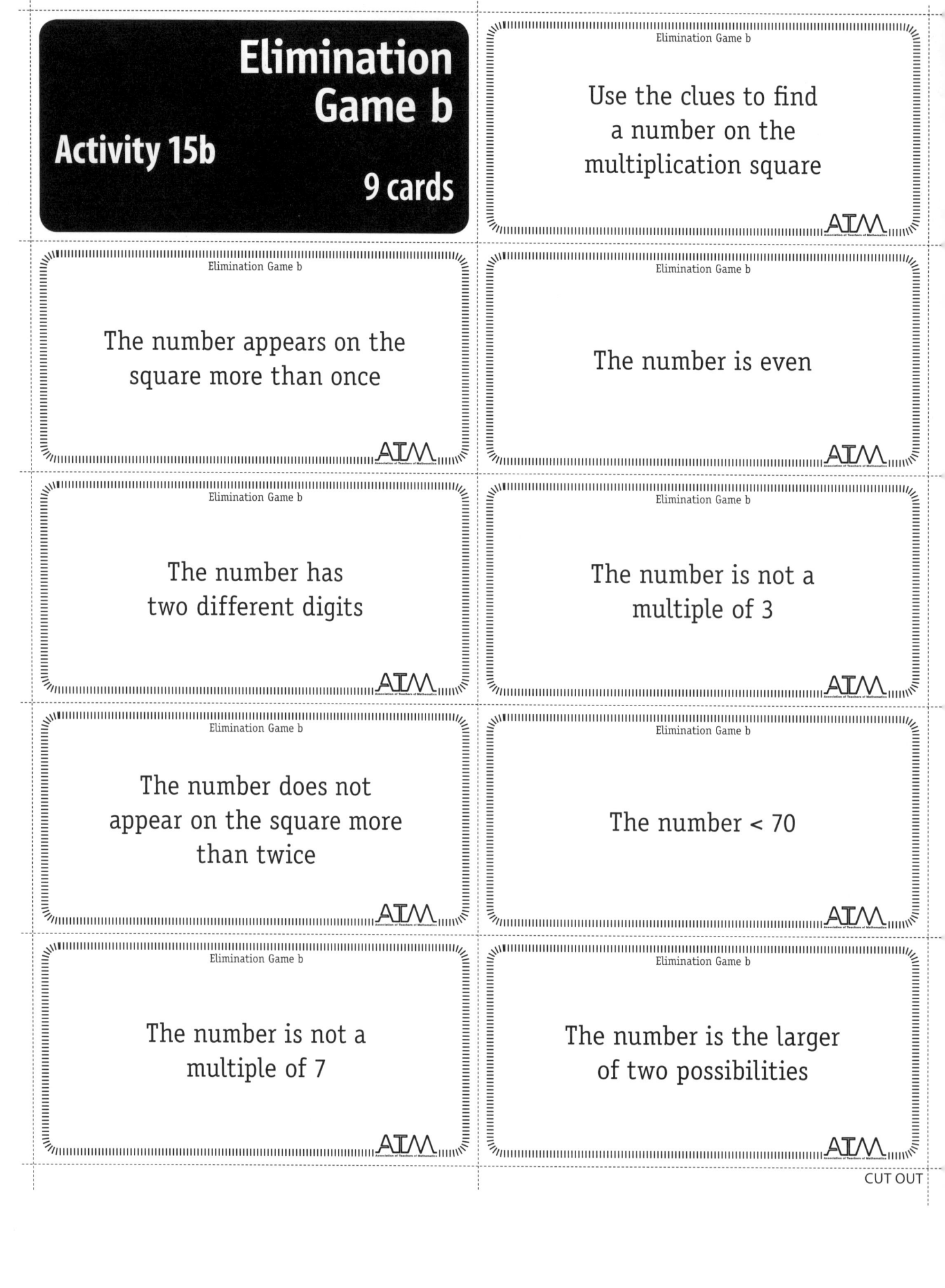

Elimination Game b

Activity 15b — 9 cards

Use the clues to find a number on the multiplication square

The number appears on the square more than once

The number is even

The number has two different digits

The number is not a multiple of 3

The number does not appear on the square more than twice

The number < 70

The number is not a multiple of 7

The number is the larger of two possibilities

Table 15b

Elimination Game b

X	1	2	3	4	5	6	7	8	9	10	11	12
1	1	2	3	4	5	6	7	8	9	10	11	12
2	2	4	6	8	10	12	14	16	18	20	22	24
3	3	6	9	12	15	18	21	24	27	30	33	36
4	4	8	12	16	20	24	28	32	36	40	44	48
5	5	10	15	20	25	30	35	40	45	50	55	60
6	6	12	18	24	30	36	42	48	54	60	66	72
7	7	14	21	28	35	42	49	56	63	70	77	84
8	8	16	24	32	40	48	56	64	72	80	88	96
9	9	18	27	36	45	54	63	72	81	90	99	108
10	10	20	30	40	50	60	70	80	90	100	110	120
11	11	22	33	44	55	66	77	88	99	110	121	132
12	12	24	36	48	60	72	84	96	108	120	132	144

Elimination Game c

Table 15c

1	2	3	4	5	6	7	8	9	10
11	12	13	14	15	16	17	18	19	20
21	22	23	24	25	26	27	28	29	30
31	32	33	34	35	36	37	38	39	40
41	42	43	44	45	46	47	48	49	50
51	52	53	54	55	56	57	58	59	60
61	62	63	64	65	66	67	68	69	70
71	72	73	74	75	76	77	78	79	80
81	82	83	84	85	86	87	88	89	90
91	92	93	94	95	96	97	98	99	100

Elimination Game C

Activity 15c

11 cards

Use the clues to find a number on the 100 square

The number is odd

The number > 35

The number < 78

The number is not prime

The number is a multiple of 3

The second digit is bigger than the first

The digits in the number are both odd

The number is not a multiple of 5

The sum of the digits in the number is 12

It is the larger of two possibilities

CUT OUT 41

Inscribed shape

Activity 16

19 cards

Inscribed shape

Draw an exact copy of a design made from two shapes

Inscribed shape

The larger shape has two sets of parallel lines

Inscribed shape

The larger shape has four right angles

Inscribed shape

The perimeter of the larger shape is 28 cm

Inscribed shape

One shape has been drawn inside the other

Inscribed shape

The smaller shape touches each of the four sides of the larger shape

---

Inscribed shape

The parts of the larger shape that are not covered by the smaller shape are blue

---

Inscribed shape

The area of the larger shape is 49cm^2

Inscribed shape

The centre of both shapes is in the same place

Inscribed shape

The smaller shape has an infinite number of axes of symmetry

Inscribed shape

The larger shape has four axes of symmetry

Inscribed shape

On the smaller shape, the length of the line across the shape that passes through the centre is the same as the length of one of the sides of the larger shape

Inscribed shape

The parts of the larger shape that are not covered by the smaller shape are congruent

Inscribed shape

When the smaller shape is placed inside the larger shape four new shapes are made

Inscribed shape

The sides of the larger shape are equal in length

Inscribed shape

The smaller shape is divided in half by a line that is perpendicular to the base of the larger shape

Inscribed shape

The left half of the smaller shape is red

Inscribed shape

The right half of the smaller shape is yellow

Inscribed shape

Both shapes are regular

CUT OUT

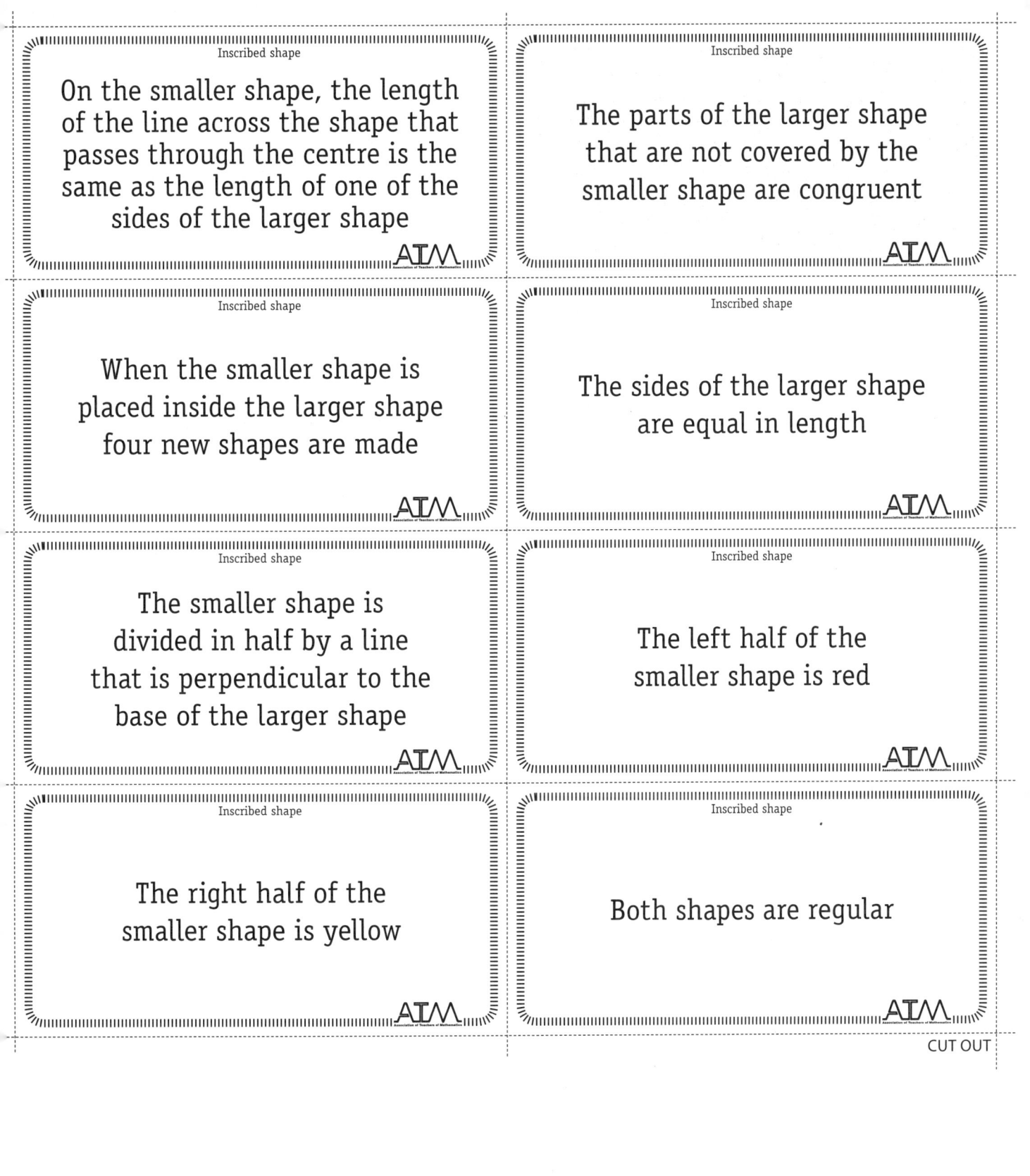

Triangle

Activity 17 — 11 cards

Find the lengths of the 3 sides of the triangle

The perimeter of the triangle is 48m

The lengths are measured in metres

The length of each side is a two-digit number

All the side lengths are larger than 10m

The ratio of the shortest side to the next largest side is 3:4

The length of the middle side plus the largest side gives an answer that is a square number

---

The length of the shortest side lies between two prime numbers

---

The triangle is not isosceles

The length of the middle side is a square number

The triangle is a right-angled triangle

The Culprit

Activity 18 — 25 cards

Find out which dog is the culprit

Mrs Brown is fed up with her dustbin being raided by a dog each week

---

She never sees the culprit

Jack knows which dog raids the dustbin but he can't describe it and he doesn't know its name

---

Jack, who is 4, is the little boy next door

Mrs Brown decides to have a line up of all the dogs in the street

---

The dogs are lined up from the biggest to the smallest

Rover is bigger than Sam

There are 7 dogs between Rover and Fido

Sam is bigger than Lucky

Beauty is standing somewhere between Sam and Fido

The Culprit

Fido is just a bit smaller than Lucky

The Culprit

There are nine dogs who live in the street

The Culprit

Jet is bigger than Lucky but smaller than Sam

The Culprit

Beauty is standing somewhere between Blackie and Ben

The Culprit

Beauty is just smaller than Sam

The Culprit

Ben is bigger than Jet

The Culprit

All the dogs are a shade of brown

The Culprit

Jack says the dog is brown

The Culprit

Jack identifies the culprit as the fifth in the line

The Culprit

Two of the dogs have patches of white on their paws

The Culprit

Charlie is the only dog between Rover and Blackie

The Culprit

Rover is the biggest

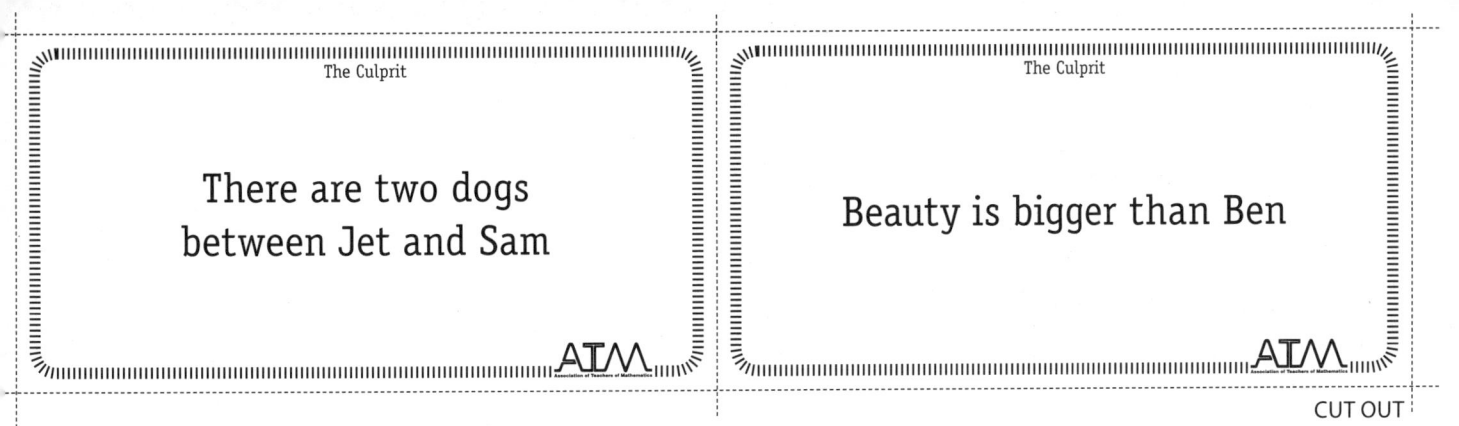

The Culprit

There are two dogs between Jet and Sam

The Culprit

Beauty is bigger than Ben

CUT OUT

47

Pie Chart

Activity 19

25 cards

Pie Chart

Use the information to build and label a pie chart

Pie Chart

The chart shows the proportion of a day that the average 11 year-old spends on different activities

---

Pie Chart

The chart should be divided into 72 sectors

Pie Chart

1 hour was spent helping at home

Pie Chart

12 different activities are included

Pie Chart

The figures were made by asking a hundred 11 year-olds, 50 boys and 50 girls to record how they spent their time over a period of one school week

---

Pie Chart

All of the times have been rounded to the nearest 20 minutes

Pie Chart

3/8 of the time was taken up with an activity we do with our eyes closed

---

Pie Chart

40 minutes was spent washing

Pie Chart

The activity that takes up most time for everybody in the survey was sleeping

---

Pie Chart

A sector is part of a circle that is enclosed by two radii and a part of the circumference

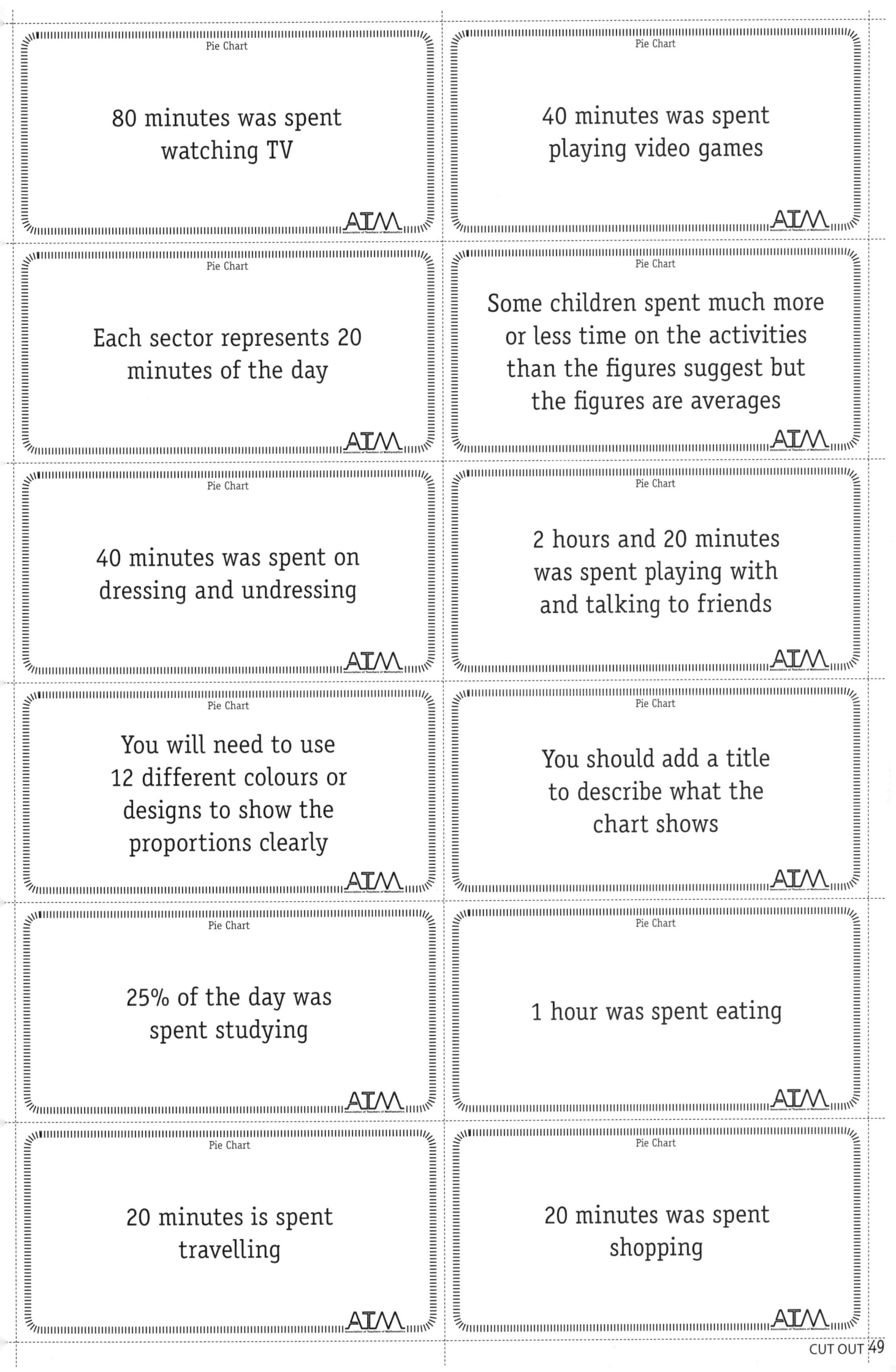

Pie Chart

80 minutes was spent watching TV

Pie Chart

40 minutes was spent playing video games

Pie Chart

Each sector represents 20 minutes of the day

Pie Chart

Some children spent much more or less time on the activities than the figures suggest but the figures are averages

Pie Chart

40 minutes was spent on dressing and undressing

Pie Chart

2 hours and 20 minutes was spent playing with and talking to friends

Pie Chart

You will need to use 12 different colours or designs to show the proportions clearly

Pie Chart

You should add a title to describe what the chart shows

Pie Chart

25% of the day was spent studying

Pie Chart

1 hour was spent eating

Pie Chart

20 minutes is spent travelling

Pie Chart

20 minutes was spent shopping

CUT OUT 49

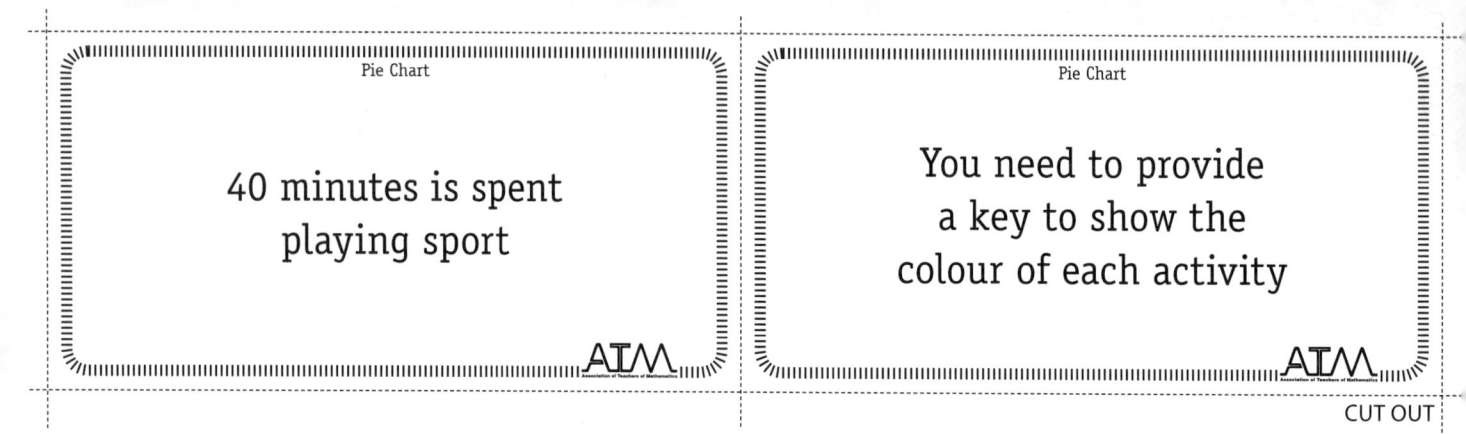

Pie Chart

40 minutes is spent
playing sport

Pie Chart

You need to provide
a key to show the
colour of each activity

Flight Times

Activity 20 — 16 cards

Flight Times

Use the information to complete the flight timetable

Flight Times

There are two flights to Oslo each day. If you miss the first flight you have to wait 8 hours and 15 minutes for the next one

---

Flight Times

Flights to Madrid leave every 6 hours

Flight Times

The last flight to Athens arrives at 00:40 local time

Flight Times

The second flight of the day to Athens leaves three hours and 25 minutes after the first

---

Flight Times

The last flight to Milan arrives at 03:20 local time

Flight Times

The second flight of the day to Milan leaves twelve hours and 25 minutes after the first

---

Flight Times

The flight to Athens takes 3 hours and 30 minutes

Flight Times

The flight to Milan takes 2 hours and 20 minutes

Flight Times

The flight to Madrid takes 10 minutes longer than the flight to Milan

---

Flight Times

The flight to Oslo takes 4 hours and 25 minutes

CUT OUT 51

Flight Times

The fourth flight of the day to Athens leaves five minutes before the last flight to Oslo

Flight Times

Madrid, Oslo and Milan are all one hour ahead of the time in London

Flight Times

The time in London is GMT (Greenwich Mean Time)

Flight Times

The time in Athens is two hours ahead of the time in London

Flight Times

The third flight of the day to Athens arrives 80 minutes before the fourth flight of the day

Flight Times

Avinunca airways Flight number	Destination	Departure time	Arrival time GMT	Arrival time Local time
AV121	Madrid	06:05		
AV411	Athens	06:40		
AV310	Oslo	06:55		
AV201	Milan	07:10		
AV412	Athens			
AV122	Madrid			
AV412	Athens			
AV413	Athens			
AV311	Oslo			
AV123	Madrid			
AV311	Athens			
AV202	Milan			
AV203	Milan			
AV124	Madrid			

CUT OUT AND LAMINATE

Who lives where?

Activity 21

28 cards

Who lives where?

Find out which family lives in each house and what is their nationality, the colour of their house, the car they own and the type of pet they have

---

Who lives where?

There are five houses in Downs Close

Who lives where?

Each house is painted a different colour

Who lives where?

The people in each house have different cars

Who lives where?

The people all have different pets

Who lives where?

Each house is lived in by families with different nationalities

---

Who lives where?

The green house is between the grey and yellow houses

Who lives where?

The Picards are French and live in the yellow house

Who lives where?

The Singh's house is at one end of the row

Who lives where?

The people in the green house have a sports car

Who lives where?

The family from Australia miss the bird life from back home and so they have built an aviary in the back garden

---

CUT OUT 53

Who lives where?

- The Smiths live in the house that is furthest from Hewlett Avenue

- The people with the pick-up truck own a cat

- The Australians own a Range Rover

- Mr Smith painted the cat flap grey to match the rest of the house

- The grey house is the furthest away from the blue house

- Mr Singh has a blue BMW which matches the colour of his house

- The Silletts live between the families from France and India

- Mr Smith lent Mr Singh his pick-up truck to carry a large new fish tank

- A dog and a cat live next door to each other

- Michelle Picard jumped into the back of the Citroen car to take her hamster to the vet

- The Red house reminds the owners of the sun and of the Australian outback which they miss

- Mr and Mrs Singh were born in Calcutta

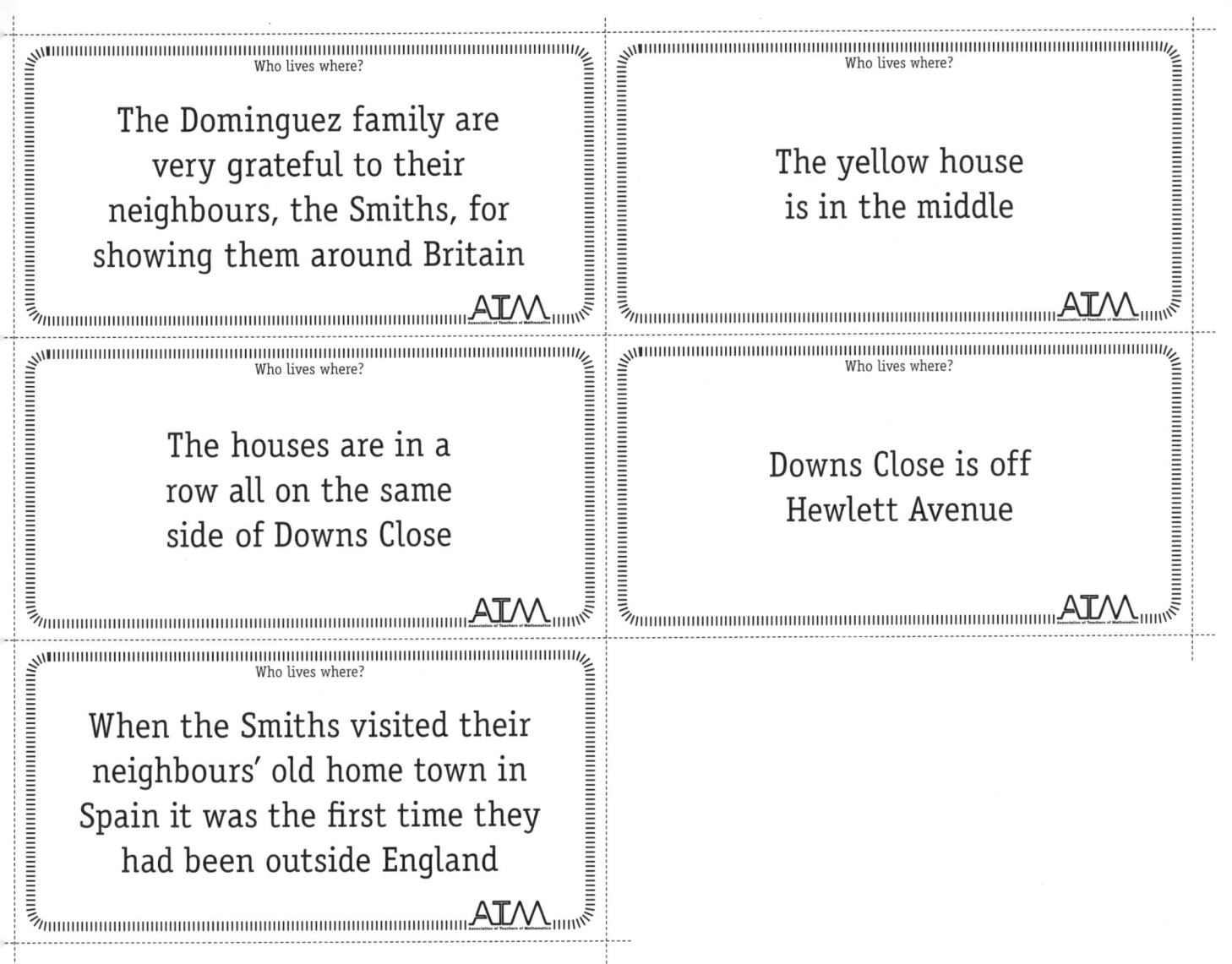

Who lives where?

The Dominguez family are very grateful to their neighbours, the Smiths, for showing them around Britain

Who lives where?

The yellow house is in the middle

Who lives where?

The houses are in a row all on the same side of Downs Close

Who lives where?

Downs Close is off Hewlett Avenue

Who lives where?

When the Smiths visited their neighbours' old home town in Spain it was the first time they had been outside England

Solid Shapes

Activity 22
22 cards

Find six different solid shapes and name them

One shape has no vertices

The shape with six faces is a regular solid

Two of the shapes are regular solids

The shape with three surfaces has two circular faces

The shape with six vertices has two triangular faces

Two of the shapes have curved surfaces

One shape has three surfaces

One shape has six vertices

The shape with five vertices has five faces

The shape with five vertices has four triangular faces

Solid Shapes

- The shape with twelve faces has pentagonal faces
- The shape with six vertices has 3 rectangular faces
- The shape with six faces has 8 vertices
- The shape with no vertices has just one surface
- One shape has two faces and one curved surface
- The shape with five vertices has one square face
- The shape with six vertices has five faces
- One shape has twelve faces
- One shape has five vertices
- One shapes has six faces
- The shape with twelve faces is a regular solid

School Trip

Activity 23

28 cards

The school day starts at 8:55

You are organising a school trip and you need to write a letter to parents telling them:
- The departure time from school
- The time of arrival back at school
- The cost of the trip for each child
- The destination
- The date of the trip
- Things the children need to take with them on the trip
- Things that they are not allowed to take

The coaches won't get back to school until after the end of the school day

---

The coaches will park opposite the school gates

There will be no need to come to school early on the day of the trip

---

The school day is 6 hours and 15 minutes long

The teachers plan to do the register in 15 minutes and then leave school immediately afterwards

---

There are 14 boys and 16 girls in the class

The difference between the contribution from the school fund and the total cost of the trip will be shared equally between all the children in the class

---

It will take about 45 minutes to drive to the Golden Valley Wildlife Park

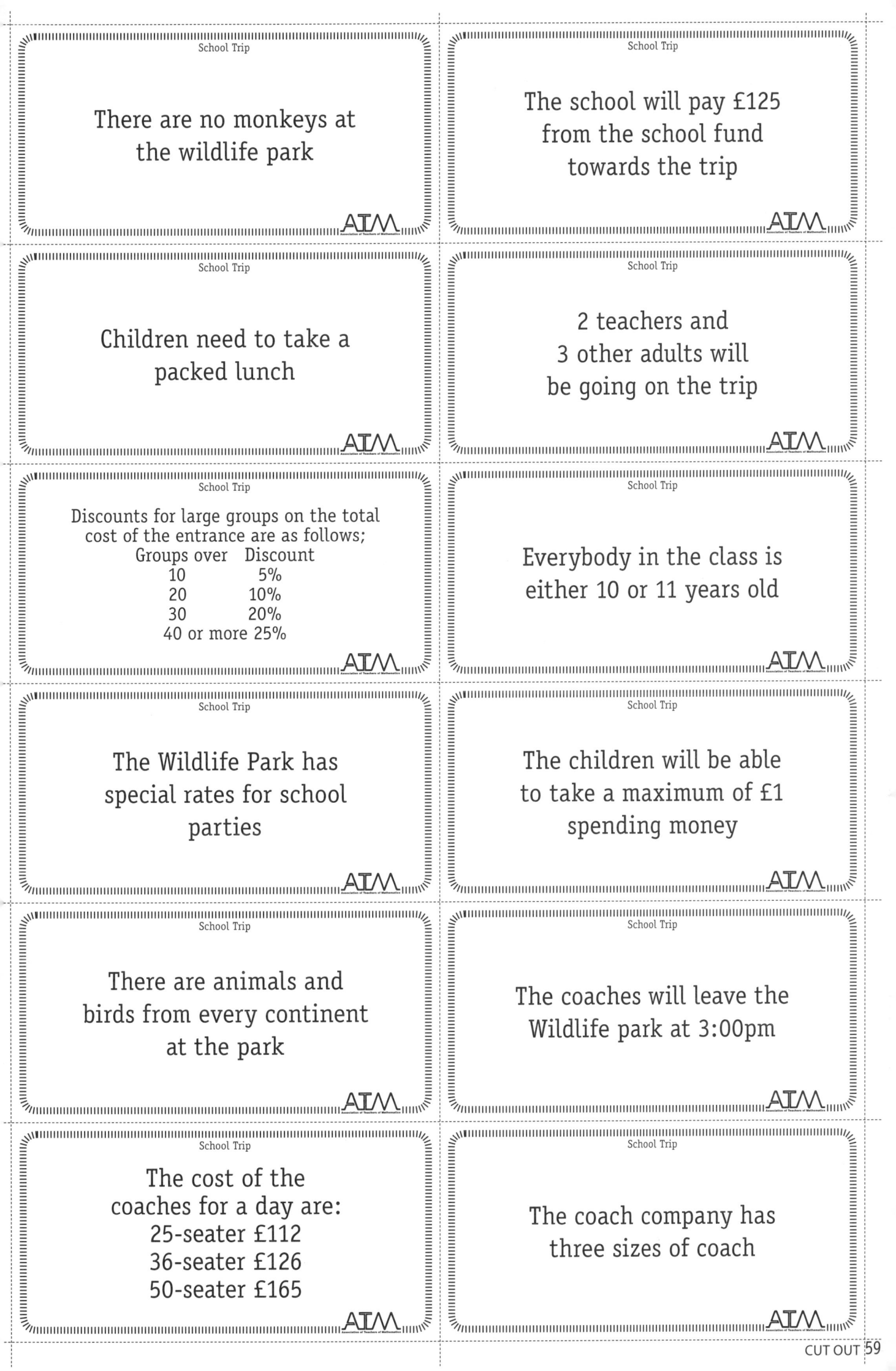

School Trip

There are no monkeys at the wildlife park

School Trip

The school will pay £125 from the school fund towards the trip

---

School Trip

Children need to take a packed lunch

School Trip

2 teachers and 3 other adults will be going on the trip

School Trip

Discounts for large groups on the total cost of the entrance are as follows;

Groups over	Discount
10	5%
20	10%
30	20%
40 or more	25%

School Trip

Everybody in the class is either 10 or 11 years old

School Trip

The Wildlife Park has special rates for school parties

School Trip

The children will be able to take a maximum of £1 spending money

---

School Trip

There are animals and birds from every continent at the park

School Trip

The coaches will leave the Wildlife park at 3:00pm

School Trip

The cost of the coaches for a day are:
25-seater £112
36-seater £126
50-seater £165

School Trip

The coach company has three sizes of coach

CUT OUT 59

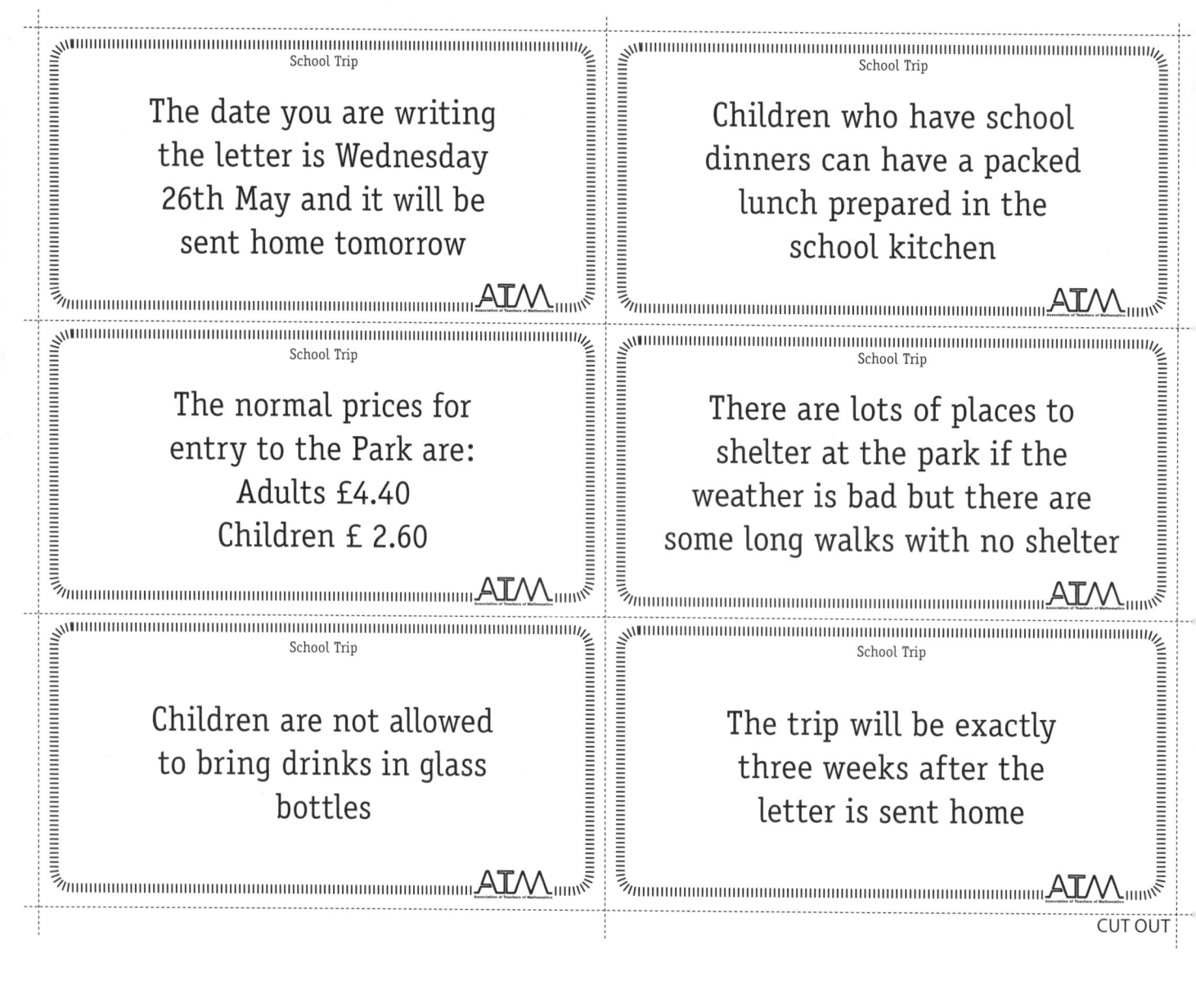

School Trip

The date you are writing the letter is Wednesday 26th May and it will be sent home tomorrow

---

School Trip

Children who have school dinners can have a packed lunch prepared in the school kitchen

---

School Trip

The normal prices for entry to the Park are:
Adults £4.40
Children £ 2.60

School Trip

There are lots of places to shelter at the park if the weather is bad but there are some long walks with no shelter

---

School Trip

Children are not allowed to bring drinks in glass bottles

School Trip

The trip will be exactly three weeks after the letter is sent home

CUT OUT

Framework shapes

Activity 24

23 cards

Framework shapes

Kits have been provided to build eight different shapes but the labels have fallen off – use the clues to work out the name and colour of each shape

---

Framework shapes

In one packet there are four green tubes that are all the same length

---

Framework shapes

With the green tubes there are four connectors with two arms

Framework shapes

In one packet there are eight yellow tubes – four are one length and four are another

---

Framework shapes

With the yellow tubes there are four connectors with three arms and one with four arms

---

Framework shapes

In one packet there are nine black tubes – there are six shorter ones that are all the same length and three longer ones of equal length

---

Framework shapes

With the black tubes there are six connectors with three arms

---

Framework shapes

In one packet there are six purple tubes – they are all different lengths

---

Framework shapes

With the purple tubes there are six connectors with two arms

Framework shapes

In one packet there are four brown tubes – two are one length and two are another

---

Framework shapes

With the brown tubes there are four connectors with two arms

CUT OUT 61

Framework shapes

- In one packet there are 12 white tubes – there are eight shorter ones that are all the same length and four longer ones of equal length

- With the white tubes there are eight connectors each with 3 arms

- The plastic tubes are joined by plastic connectors

- Different colour strips have been used for each shape

- Each shape is made by joining plastic tubes together

- There are different connectors some have two arms, some three and one has four

- In one packet there are twelve blue tubes that are all the same length

- With the blue tubes there are eight connectors with three arms

- 4 of the shapes are 2D shapes

- 4 of the shapes are 3D solids

- In one packet there are three red tubes that are all the same length

- With the red tubes there are three connectors with two arms

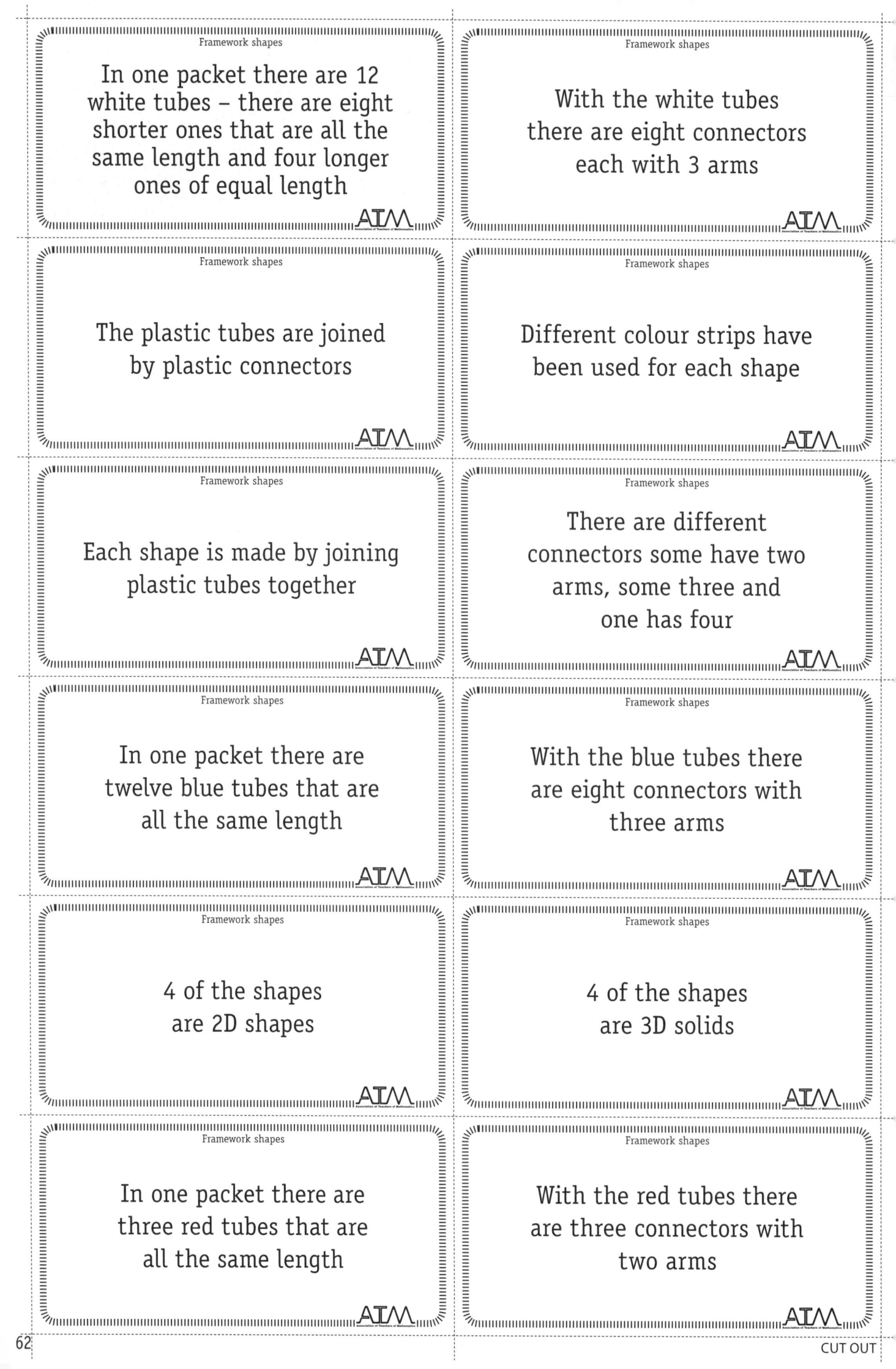

Bookshop

Activity 25 — 33 cards

Bookshop

Use the clues to work out the prices you will charge for 10 new books you are going to sell in your bookshop

---

Bookshop

You have ordered five copies of Dave Brush's book

Bookshop

Norman Allsop and the Boring Afternoon
– Arthur Dullard
Children's Fiction
Wholesale price £ 2.00

Bookshop

Two Left Feet:
The Story of Darren Dash
Autobiography
Wholesale price: £5.00

Bookshop

All the Greenhouse Gang books have been best-sellers

Bookshop

One cookery book is likely to sell more quickly than the other because it comes from a TV programme, so you order more of that one

---

Bookshop

Oliphant's Gravy Yard is a popular TV programme

Bookshop

Grow Bigger Tomatoes
– Bert Compost
Non fiction   Gardening
Wholesale price: £8.00

Bookshop

The Daniel Lambert Diet:
How to Stay Overweight
Daniel Lambert   Non-fiction
Health and fitness
Wholesale price: £5.50

Bookshop

The books by Amanda Volebotherer and Darren Dash will be your special offers for this week

---

Bookshop

You are not sure that diet books written for people who want to stay overweight will sell well, so you order just four copies

---

CUT OUT 63

Bookshop

The Greenhouse Gang Strike Out – Amanda Volebotherer
Children's Fiction
Wholesale price £3.50

Bookshop

You order 24 copies of the autobiography because the author has just become captain of the England football team and married the Duchess of York

---

Bookshop

Gardening books sell quite well, so you buy 6 copies

Bookshop

100 Meals with Beans and Cabbage - Winn D Knight
Non-fiction Cookery
Wholesale price £4.50

Bookshop

1001 daft jokes – edited by Dave Brush
Children's humour
Wholesale price: £ 2.40

Bookshop

For special offers you take 25% off the normal sale price

Bookshop

George T Cupp has only just recovered from a nasty accident that involved a dishwasher

---

Bookshop

When Amanda Volebotherer is not writing she works as a gamekeeper

---

Bookshop

Dave Brush is about to get married to the famous Chinese actress Barbara Lu. After they are married she plans to keep her maiden name by making a double-barrelled surname with Dave's

---

Bookshop

Beastly Basil and the Aunt Invasion - George T Cupp
Children's fiction
Wholesale price: £2:20

Bookshop

The method of finding the sale price only changes when you have special offers

---

Bookshop

You have ordered 10 copies of one cookery book and 2 copies of the other

Bookshop

Darren Dash was offered a ghostwriter to help him with his book but he was too scared to take up the offer. In the end his mum helped him. She did the writing whilst Darren coloured in the cover

Bookshop

You have ordered 40 copies of Amanda Volebotherer's new book

Bookshop

Your bookshop buys books from a wholesale supplier

Bookshop

To work out the sale price you add something to the price you pay the supplier so that you can cover the costs of running your shop and make some profit

Bookshop

You have ordered 10 copies of all the children's fiction books except one

Bookshop

The suppliers give a discount on the wholesale price based on the number of copies you order.

Number of books	Discount
1-4	0%
5-9	10%
10-19	25%
20-29	30%
30-39	35%
40+	40%

Bookshop

Cookery Without Tears
– Barney Oliphant

Non-fiction   Cookery

Wholesale price: £6.40

Bookshop

Showdown at the Slug Rodeo – Ben D Legg

Children's Fiction

Wholesale price: £2.80

Bookshop

Bert Compost's real name is Gavin Fotherington-Thomas

Bookshop

You take the wholesale price of the book, after the discount has been taken off, and then double that price to get the sale price

CUT OUT

65

Answers 1–6

1 The great race
Order of finishing:
Red
Blue
Green
Orange
Yellow
Pink
Black
White
Purple

2 Shape arrangement

Row 1: Red Triangle, Yellow Square, Blue Circle, Green Hexagon
Row 2: Green Hexagon, Blue Circle, Yellow Square, Red Triangle

3 Football matches — Results:
Chelsea 2 Spurs 1   Att: 39 000
Liverpool 4 Southampton 1 Att: 45 000
Man Utd 0 Arsenal 3 Att: 59 000
Wolves 1 Newcastle 1 Att: 36 000

4 What do we know 1 - Trip to London
1. Yes we can answer (born 29-04-30)
2. We can't answer accurately. John spent up to £4.75. We need to know the cost of a bacon sandwich

4 What do we know 1 - continued
3. We can't answer this
4. Yes we can answer - 1 hour and 25 minutes
5. We can't answer this
6. Yes, he spent £56
7. We can't answer accurately but it must have been about 11:00am because we know he left at 8:48, the train took 1 hour and 25 minutes and the taxi ride took 35 minutes and he arrived just in time
8. Yes, he saved £14
9. We can't answer this except to say he lives about 1 hour and 15 minutes away by train
10. We can't answer this

5 Guess my number
1. 42
2. 39
3. 73

6 Summer holiday
£2068

Answers

7 – 9

7 Mystery symbols – A

$	=	1
★	=	2
€	=	3
£	=	4
♣	=	5
♠	=	6
e	=	7

Mystery symbols – B

□	=	1
✲	=	2
⌬	=	3
✎	=	4
◆	=	5
▲	=	6
Ω	=	7
✾	=	8
♥	=	9

8 Chocolate box

1. Truffle Delight
2. Coconut Heaven
3. Turkish Delight
4. Strawberry Fizz
5. Creamy Caramel
6. Lemon Twist
7. Tantalising Cream
8. Liquid Cherry
9. Chocolate Brazil
10. Hazelnut Crackle

9 Triangle Construction

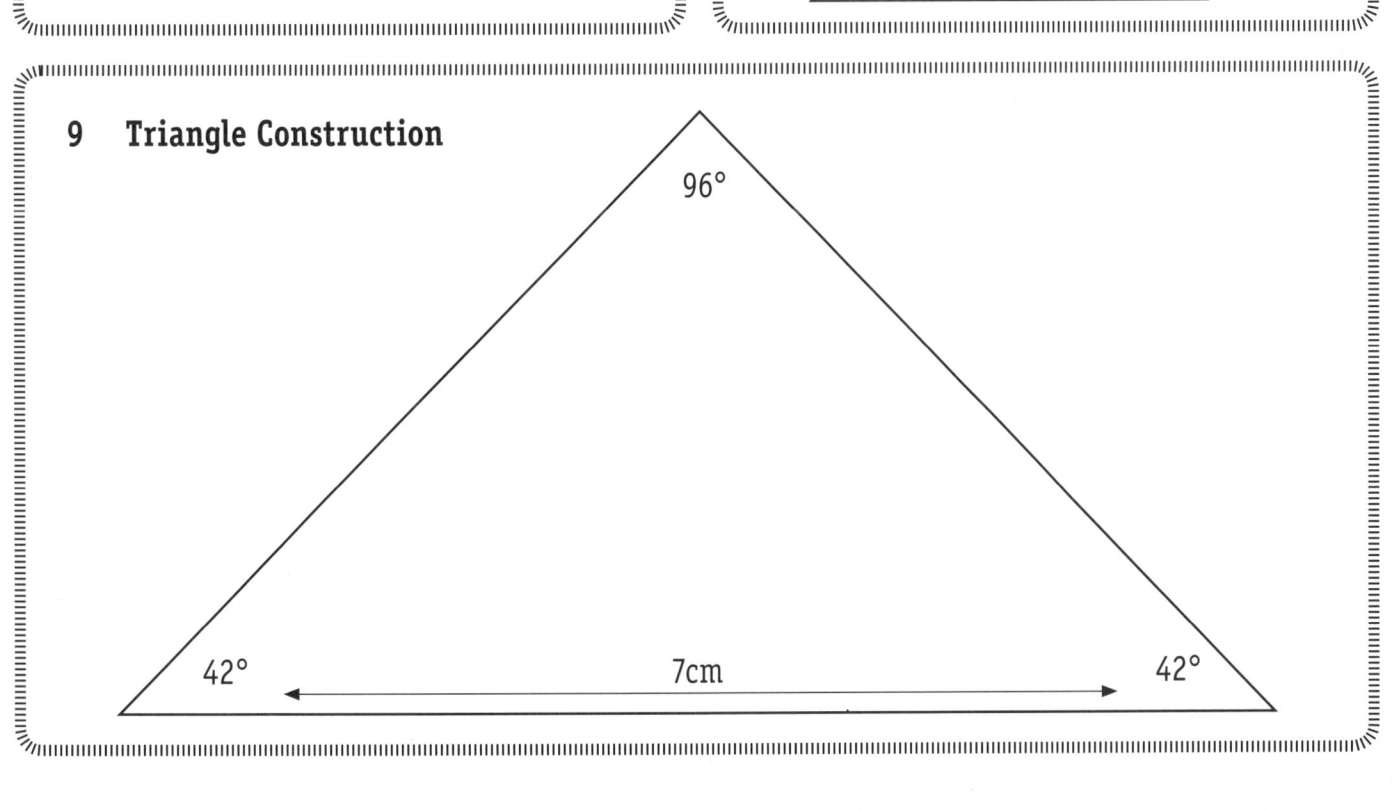

96°, 42°, 7cm, 42°

Answers

10 – 12

10 Birthday presents
Bracelet - Cone
Harry Potter book - Cuboid
Maltesers - Cylinder
Jumper - Cube
Bubble bath - Sphere
Hair decorations - Hexagonal prism

11 Cake shop
Cake A - Chelsea bun   50p
Cake B - Doughnut   35p
Cake C - Hot cross bun 25p
Cake D - Éclair   30p
Cake E - Swiss roll   40p

12 Who sits where?
1. Holly
2. Deepak
3. Diego
4. Louisa
5. Alex
6. Harshini
7. George
8. Patrick
9. Sophie
10. Julie
11. Ahmed
12. Ben
13. Eloise
14. Anita
15. Jim
16. Megan
17. Catalina
18. Ranjit
19. Barry
20. Satvia

Answers

13 – 18

13 What do we know 2 - Visit to the zoo
1. Yes we can answer 11:45
2. We can't answer accurately but we know that the chimpanzee was popular.
3. We can't answer this.
4. Yes, there will be 4 adults (don't forget the driver)
5. Yes, they will pay £2.30
6. Yes, they would be 9 spare seats
7. Yes, they will pay £69
8. We can't answer because we only know how long it takes to get there.
9. Yes, they would have paid £68.20
10. Yes, they arrived at 11:15

14 Queue for the Mirror Maze
1. Sally
2. John
3. Grace
4. Mike
5. Pippa
6. James
7. Sarah
8. Andrew
9. Satvia
10. Ahmed

15 Elimination games
1. 85p
2. 50
3. 57

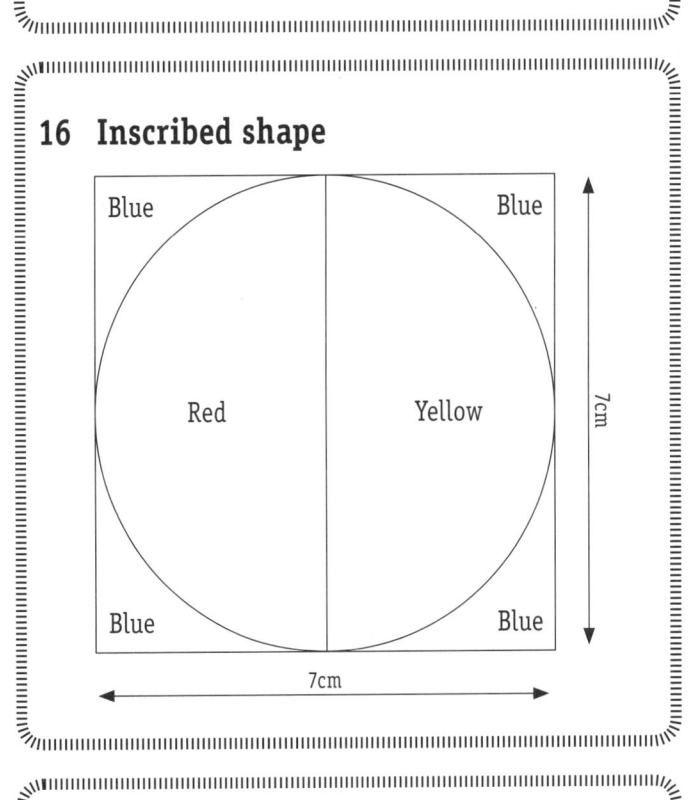

16 Inscribed shape

Square 7cm × 7cm with inscribed circle. Corners labelled Blue (top-left), Blue (top-right), Blue (bottom-left), Blue (bottom-right). Circle divided vertically into Red (left) and Yellow (right).

17 Triangle
12m, 16m and 20m

18 The culprit
Beauty

Answers

19 – 21

19 Pie chart

Number of sectors for each activity:

Sleeping   -   27

Playing video games   -   2

Dressing and undressing   -   2

Eating   -   3

Watching TV   -   4

Sport   -   2

Studying   -   18

Washing   -   2

Playing with a talking to friends   -   7

Shopping   -   1

Helping at home   -   3

Travelling   -   1

21 Who lives where?

Downs Close

| Family name: Smith |
| Pet: Cat |
| House colour: Grey |
| Car: Pick-up truck |
| Nationality: British |

| Family name: Dominguez |
| Pet: Dog |
| House colour: Green |
| Car: Sports car |
| Nationality: Spanish |

| Family name: Picard |
| Pet: Hamster |
| House colour: Yellow |
| Car: Citroen |
| Nationality: French |

| Family name: Sillet |
| Pet: Birds |
| House colour: Red |
| Car: Range Rover |
| Nationality: Australian |

| Family name: Singh |
| Pet: Fish |
| House colour: Blue |
| Car: BMW |
| Nationality: Indian |

Hewlett Avenue

20 Flight times

Avinunca airways Flight number	Destination	Departure time	Arrival time GMT	Arrival time Local time
AV121	Madrid	06:05	08:35	09:35
AV411	Athens	06:40	10:10	12:10
AV310	Oslo	06:55	11:20	12:20
AV201	Milan	07:10	09:30	10:30
AV412	Athens	10:05	13:35	15:35
AV122	Madrid	12:05	14:35	15:35
AV412	Athens	13:45	17:15	19:15
AV413	Athens	15:05	18:35	20:35
AV311	Oslo	15:10	19:35	20:35
AV123	Madrid	18:05	20:35	21:35
AV311	Athens	19:10	22:40	00:40
AV202	Milan	19:35	21:55	22:55
AV203	Milan	00:00	02:20	03:20
AV124	Madrid	00:05	02:35	03:35

Answers

22 – 25

22 Solid shapes
Cube
Triangular prism
Sphere
Cylinder
Square-based Pyramid
Dodecahedron

23 School trip
The time of departure from school-
9:10am

The time of arrival back at school-
3:45pm

The contribution towards the cost of the trip from each child-
£2.70

Where you are going-
Golden Valley Wildlife Park

The date of the trip-
Thursday June 17th

Things the children need to take with them on the trip:
Packed lunch
No more than £1 spending money
A waterproof coat or jacket

Things that they are not allowed to take-
Drinks in glass bottles

24 Framework shapes
Cube - Blue
Equilateral Triangle - Red
Square - Green
Square-based Pyramid - Yellow
Triangular prism - Black
Cuboid - White
Irregular Hexagon - Purple
Rectangle - Brown

25 Book shop - *Prices:*
The Daniel Lambert diet: How to Stay Overweight - Daniel Lambert **£11.00**

1001 daft jokes - edited by Dave Brush **£4.32**

Grow Bigger Tomatoes - Bert Compost **£14.40**

Beastly Basil and the Invasion of the Aunts - **£3.30** George T Cupp

Two Left Feet: The Story of Darren Dash – Darren and Phyllis Dash **£5.25**

Norman Allsop and the Boring Afternoon – Arthur Dullard **£3:00**

Showdown at the Slug Rodeo – Ben D Legg **£4.20**

Cookery without Tears – Barney Oliphant **£9.60**

The Greenhouse Gang Strike Out – Amanda Volebotherer **£3.15**

100 meals with Beans and Cabbage - Winn D Knight **£9.00**